AF882131

nephron Journals

Founded 1964 by G. Richet and G.E. Schreiner
Former Editors: S. Giovannetti (1972–1998); G.M. Berlyne (1972–2002); S. Ito (1998–2002)

KARGER

Printed in Switzerland on acid-free paper by Reinhardt Druck, Basel

Appearance: 3 volumes per year (12 issues)

Guidelines for Authors

nephron
Journals

Submission

Manuscripts for
Nephron Clinical Practice
Nephron Experimental Nephrology
Nephron Physiology
should be submitted online at

www.karger.com/nef

Should you experience problems with your submission, please contact

Editorial Office 'Nephron'
Royal Free & University College Medical School
University College London
Gower Street
London WC1E 6BT (UK)
Tel. +44 20 7679 0886
Fax +44 20 7679 0897
E-Mail nephron@ucl.ac.uk

If a specific subjournal is not designated, or an article would be more suitable for a subjournal other than that selected by the authors, the Editorial Office will assign it to the most appropriate of the 3 'Nephron' journals.

Minireviews: Are to be submitted only by invitation from the Editor, although prospective authors are welcome to contact the Editor to discuss whether a topic might suitable for a Minireview via nephron@ucl.ac.uk.

Suggested reviewers: Please list four complete mailing and e-mail addresses of potential reviewers for your manuscript. The names you suggest should not be 'Nephron' Editorial Board members; should not have been co-authors or co-workers over the past 3 years nor should they be persons working in the same institution as yourself or any of the co-authors.

Conditions

All manuscripts are subject to independent and editorial review. Manuscripts concerning clinical trials will only be considered if the trial has been fully registered (see page opposit). Manuscripts are received with the explicit understanding that they are not under simultaneous consideration by any other publication. Submission of an article for publication implies transfer of the copyright from the author to the publisher upon acceptance. Accepted papers become the permanent property of 'Nephron' journals and may not be reproduced by any means, in whole or in part, without the written consent of the publisher. It is the author's responsibility to obtain permission to reproduce illustrations, tables, etc. from other publications.

Arrangement

All pages should be numbered consecutively and carry the author's name in the upper right corner.

Title page: The first page of each paper should indicate the title, the authors' names, the institute where the work was conducted, and a short title for use as running head.
NB: Authors wishing to preserve the phonetic meaning of diacritics (PubMed reduces diacritics to their root characters) must spell their names accordingly when submitting manuscripts (e.g. Müller should be Mueller).

Full address: The exact postal address of the corresponding author complete with postal code must be given at the bottom of the title page. Please also supply phone and fax numbers, as well as e-mail address.

Units: 'Nephron' journals requires that all articles be submitted with data in SI units and gravimetric units.

Key words: For indexing purposes, a list of 3–10 key words in English is essential.

Abstract: Each paper needs an abstract of up to 200 words. It should be structured as follows:

Background/Aims:	What is the major problem that prompted the study?
Methods:	How was the study carried out?
Results:	Most important findings?
Conclusion:	Most important conclusion?

Footnotes: Avoid footnotes.

Tables and illustrations: Tables are part of the text. Place them at the end of the text file. Illustration data must be stored as separate files. Do not integrate figures into the text. Electronically submitted b/w half-tone and color illustrations must have a final resolution of 300 dpi after scaling, line drawings one of 800–1,200 dpi.

Color illustrations
Online edition: Color illustrations are reproduced free of charge. In the print version, the illustrations are reproduced in black and white. Please avoid referring to the colors in the text and figure legends.
Print edition: Up to 6 color illustrations per page can be integrated within the text at CHF/USD 760.00 per page.

References: In the text identify references by Arabic numerals [in square brackets]. Material submitted for publication but not yet accepted should be noted as 'unpublished data' and not be included in the reference list. The list of references should include only those publications which are cited in the text. Number references in the order in which they are first mentioned in the text; do not list alphabetically. The surnames of the authors followed by initials should be given. There should be no punctuation other than a comma to separate the authors. Preferably, please cite all authors. Abbreviate journal names according to the Index Medicus system. (Also see International Committee of Medical Journal Editors: Uniform requirements for manuscripts submitted to biomedical journals. N Engl J Med 1997;336:309–315. www.icmje.org).

Digital Object Identifier (DOI)
S. Karger Publishers supports DOIs as unique identifiers for articles. A DOI number will be printed on the title page of each article. DOIs can be useful in the future for identifying and citing articles published online without volume or issue information. More information can be found at www.doi.org.

Examples
(a) Papers published in periodicals: Tomson C: Vascular calcification in chronic renal failure. Nephron Clin Pract 2003;93:c124–c130.
(b) Papers published only with DOI numbers: Theoharides TC, Boucher W, Spear K: Serum interleukin-6 reflects disease severity and osteoporosis in mastocytosis patients. Int Arch Allergy Immunol DOI: 10.1159/000063858.
(c) Monographs: Matthews DE, Farewell VT: Using and Understanding Medical Statistics, ed 3, revised. Basel, Karger, 1996.
(d) Edited books: Kashihara N, Sugiyama H, Makino H: Implication of apoptosis in progression of renal diseases; in Razzaque MS, Taguchi T (eds): Renal Fibrosis. Contrib Nephrol. Basel, Karger, 2003, vol 139, pp 156–172.

Author's Choice™

With this option the author can choose to make his article freely available online against a one-time fee of USD 2'500.00. This fee is independent of any standard charges for supplementary pages, color images etc. which may apply. More information can be found at www.karger.com/authors_choice.

Page Charges

There are no page charges for papers of 4 or fewer printed pages (including tables, illustrations and references). Each additional complete or partial page is charged to the author at CHF/USD 290.00. Four printed pages are equal to approx. 12 manuscript pages (including tables, illustrations and references).

Proofs

Unless indicated otherwise, proofs are sent to the first-named author and should be returned with the least possible delay. Alterations made in proofs, other than the correction of printer's errors, are charged to the author.

Reprints

Order forms and a price list are sent with the proofs. Orders submitted after the issue is printed are subject to considerably higher prices.

KARGER

Fax +41 61 306 12 34
E-Mail karger@karger.ch
www.karger.com

The Guidelines for Authors are available at:
www.karger.com/nef_Guidelines

A Uniform Clinical Trial Registration Policy for Journals of Kidney Diseases, Dialysis and Transplantation

Clinical trials provide important information that can profoundly influence the direction of medical research and clinical care. It is crucial that this process is not unduly influenced by selective publication of clinical trial results. Information regarding all clinical trials should therefore be made available to the general public.

In common with other nephrology journals, *Nephron* requires that from 2006 all submitted manuscripts concerning clinical trials must be registered in a public trials registry.

The full text of a recent statement published by the editors of: *Kidney International; The Journal of the American Society of Nephrology; Nephrology, Dialysis, Transplantation; The American Journal of Transplantation; The American Journal of Kidney Diseases,* and *Transplantation* is reproduced below. The policy described in this text has now been formally adopted by *Nephron.*

'Medical research can be seriously compromised by the selective publication of clinical trial results. Therefore, it is imperative that information regarding clinical trials should be available to the general public.

We require, as a condition of consideration for publication, registration in a public trials registry. Trials must register at or before the onset of patient enrolment.' Since 1 April 2006, *Nephron* no longer accepts manuscripts concerning non-registered trials.

'We define a clinical trial as any research project that prospectively assigns human subjects to intervention or comparison groups to study the cause-and-effect relationship between a medical intervention and a health outcome. Studies designed for other purposes, such as to study pharmacokinetics or major toxicity (e.g. phase 1 trials) are exempt.

We do not advocate any one particular registry, but registration must be with a registry that meets the following minimum criteria:

- Accessible to the public at no charge.
- Searchable by standard, electronic (Internet-based) methods.
- Open to all prospective registrants free of charge or at minimal cost.
- Validates registered information.
- Identifies trials with a unique number.
- Includes information on the investigator(s), research question or hypothesis, methodology, intervention and comparisons, eligibility criteria, primary and secondary outcomes measured, date of registration, anticipated or actual start date, anticipated or actual date of last follow-up, target number of subjects, status (anticipated, ongoing or closed) and funding source(s).

Registries that currently meet these criteria include:

- The registry sponsored by the US National Library of Medicine www.clinicaltrials.gov
- The International Standard Randomised Controlled Trial Number registry www.controlled-trials.com
- The Cochrane Renal Group registry www.cochrane-renal.org/trialsubmissionform.php
- The National (UK) Research Register www.update-software.com/national/
- European Clinical Trials Database http://eudract.emea.eu.int/

KARGER

Fax +41 61 306 12 34
E-Mail karger@karger.ch
www.karger.com

The Guidelines for Authors are available at:
www.karger.com/nef_Guidelines

General Information

nephron
Journals

ISSN Print Edition: 0028–2766
ISSN Online Edition: 1423–0186

Journal Homepage: www.karger.com/nef

Publication Data: 'Nephron' is published 12 times a year. Volumes 105–107, each with 4 issues, appear in 2007.

Subscription Rates: Subscriptions run for a full calendar year. Prices are given per year.

Personal subscription:

Print or Online	Print+Online combined
CHF 848.25	CHF 983.25
EUR 628.50	EUR 727.50
USD 771.00	USD 894.00

postage and handling (added to print and print+online)
CHF 69.00 Europe, CHF 102.60 Overseas
EUR 49.80
USD 94.20

Institutional subscription:

Print or Online	Print+Online combined
CHF 3393.00	CHF 3732.00
EUR 2514.00	EUR 2763.00
USD 3084.00	USD 3393.00

postage and handling (added to print and print+online)
CHF 86.40 Europe, CHF 128.40 Overseas
EUR 62.40
USD 117.60

Airmail surcharge: CHF 87.00 / USD 78.60

Online edition of each subjournal is available separately:

Clinical Practice *ISSN 1660–2110*
CHF 330.00, EUR 244.00, USD 300.00
Experimental Nephrology *ISSN 1660–2129*
CHF 330.00, EUR 244.00, USD 300.00
Physiology *ISSN 1660–2137*
CHF 330.00, EUR 244.00, USD 300.00

Discount subscription prices:
International Society of Nephrology;
American Society of Nephrology.

Back Volumes and Single Issues: Information on availability and prices of single print issues and print or electronic back volumes can be obtained from Customer Service at *service@karger.ch.*

Bibliographic Indices: This journal is regularly listed in bibliographic services, including *Current Contents®* and PubMed/MEDLINE.

Photocopying: This journal has been registered with the Copyright Clearance Center (CCC), as indicated by the code appearing on the first page of each article. For readers in the US, this code signals consent for copying of articles for personal or internal use, or for the personal or internal use of specific clients, provided that the stated fee is paid per copy directly to
Copyright Clearance Center Inc.
222 Rosewood Drive
Danvers, MA 01923 (USA)
A copy of the first page of the article must accompany payment. Consent does not extend to copying for general distribution, for promotion, for creating new works, or for resale. In these cases, specific written permission must be obtained from the copyright owner,
S. Karger AG, P.O. Box
CH–4009 Basel (Switzerland).

Subscription Orders:
Orders can be placed at agencies, bookstores, directly with the Publisher

S. Karger AG
Medical and Scientific Publishers
P.O. Box
CH–4009 Basel
Switzerland
(for courier services only:
Allschwilerstrasse 10
CH–4055 Basel)
Tel. +41 61 306 11 11
Fax +41 61 306 12 34
E-Mail karger@karger.ch
www.karger.com

or further Karger offices or representatives:

France:
Librairie Medi-Sciences Sarl
36, bd de Latour-Maubourg
75007 Paris
France
Tél. +33 (0) 1 45 51 42 58
Fax +33 (0) 1 45 56 07 80
E-Mail librairie@medi-sciences.fr
www.medi-sciences.fr

Germany:
S. Karger GmbH
Postfach
79095 Freiburg
Deutschland
(Hausadresse: Lörracher Strasse 16A
79115 Freiburg)
Tel. +49 761 45 20 70
Fax +49 761 45 20 714
E-Post information@karger.de
www.karger.de

India, Bangladesh, Sri Lanka:
Panther Publishers Private Ltd.
33, First Main
Koramangala First Block
Bangalore 560 034
India
Tel. +91 80 25505 836
Tel. +91 80 25505 837
Fax +91 80 25505 981
E-Mail panther_publishers@vsnl.com
www.pantherpublishers.com

Japan:
Karger Japan, Inc.
Yushima S Bld. 3F
4-2-3, Yushima, Bunkyo-ku
Tokyo 113-0034
Japan
Tel. +81 3 3815 1800
Fax +81 3 3815 1802
E-Mail publisher@karger.jp

People's Republic of China:
Creative Consulting (Shanghai) Co., Ltd.
1 Fuxing Zhong Lu
Suite 1208
Shanghai 200021
China
Tel. (86 21) 63 91 92 90
Fax (86 21) 63 91 92 95
E-Mail karger_china@yahoo.com.cn

South America and Central America:
Cranbury International LLC
7 Clarendon Ave., Suite 2
Montpelier, VT 05602
USA
Tel. +1 802 223 6565
Fax +1 802 223 6824
E-Mail
eatkin@cranburyinternational.com
www.cranburyinternational.com

United Kingdom, Ireland:
S. Karger AG
c/o London Liaison Office
4 Rickett Street
London SW6 1RU
United Kingdom
Tel. +44 (0) 20 7386 0500
Fax +44 (0) 20 7610 3337
E-Mail uk@karger.ch

USA:
S. Karger Publishers, Inc.
26 West Avon Road
P.O. Box 529
Farmington, CT 06085
USA
Toll free: +1 800 828 5479
Tel. +1 860 675-7834
Fax +1 860 675-7302
E-Mail karger@snet.net

Change of Address:
Both old and new address should be sent to the subscription source.

KARGER

Fax +41 61 306 12 34
E-Mail karger@karger.ch
www.karger.com

The Journal Home Page is available at:
www.karger.com/nef

A concise assessment of current knowledge and future developments

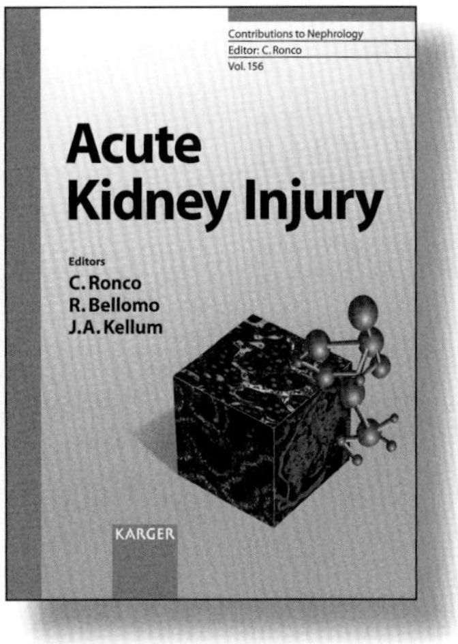

Acute Kidney Injury

Editors
Claudio Ronco
Rinaldo Bellomo
John A. Kellum

Nephrology; Internal Medicine; Dialysis, Cell Biology

Main Headings

- Critical Care Nephrology Issues
- Epidemiology and Pathogenesis of AKI, Sepsis and MOF
- Evaluation of Illness Severity
- Metabolism, Electrolytes and Acid-Base Disorders
- Nursing Issues in Critical Care Nephrology
- Early Diagnosis and Prevention of AKI
- Practice Patterns for RRT in the ICU
- Which Treatment for AKI in the ICU
- Extracorporeal Treatment for Specific Indications
- New Trials and Meta-Analyses

Acute kidney injury is defined as an abrupt change in serum creatinine and/or urine output, and a majority of patients admitted to the ICU have some evidence of the disorder. Unfortunately, treatment for this complex syndrome is as yet lacking and understanding is limited.
An interdisciplinary panel of experts has contributed to this volume, illuminating some of the fundamental and complex aspects of the disorder ranging from pathophysiology to treatment, from emerging biomarkers to genetic polymorphisms. Other contributions focus on immunological issues or the many complications of acute kidney injury and co-morbid conditions encountered, covering the fundamentals as well as the latest developments. Moreover, important technical aspects of extracorporeal therapies including vascular access, anticoagulation or fluid composition are introduced, and different approaches to renal support from intermittent dialysis to continuous therapies and hybrid techniques are discussed. A description of advanced extracorporeal techniques of organ support and their role in the management of sepsis and acute kidney injury in the context of an overall strategy of multi-organ failure management concludes the discussions.
This volume not only provides a practical and up-to-date summary of current knowledge and technology, but also imparts a fundamental understanding of the pathogenesis and likely future developments in this field. It also serves to challenge and re-examine the fundamental underlying assumptions we hold regarding critical illness in general and acute kidney injury in particular.

KI07652

www.karger.com/conep

Contributions to Nephrology, Vol. 156
Series Editor: Ronco, C. (Vicenza)
ISSN 0302–5144

Acute Kidney Injury
Editors: Ronco, C. (Vicenza); Bellomo, R. (Melbourne, Vic.); Kellum, J. A. (Pittsburgh, PA)
XII + 464 p., 63 fig., 2 in color, 40 tab., hard cover, 2007
CHF 286.– / EUR 204.50 / USD 260.00
Prices subject to change
EUR price for Germany, USD price for USA only
ISBN 978–3–8055–8271–1

KARGER

Order Form

Please send: ____ copy/ies

Postage and handling free with prepayment

Payment:
Please charge to my credit card
☐ American Express ☐ Diners ☐ Eurocard
☐ MasterCard ☐ Visa

Card No.: ____________

Exp. date: ____________

CVV/CVC ____________
(3 digits in the signature field on the back of VISA and Mastercard)

☐ Check enclosed ☐ Please bill me

Orders may be placed with any bookshop, subscription agency, directly with the publisher or through a Karger distributor.

Fax: +41 61 306 12 34

S. Karger AG, P.O. Box, CH–4009 Basel (Switzerland)
E-Mail orders@karger.ch, **www.karger.com**

Name/Address: ____________

____

____

____

____

Date: ____________

Signature: ____________

nephron
Journals

Glomerular Proteinuria and Hematuria

Crossing the Bridge from Molecules to Man

Guest Editors
Peter W. Mathieson, Bristol, UK
Jeffrey H. Miner, St. Louis, Mo.

13 figures, 1 in color, and 3 tables, 2007

KARGER
Basel · Freiburg · Paris · London · New York ·
Bangalore · Bangkok · Singapore · Tokyo · Sydney

S. Karger
Medical and Scientific Publishers
Basel • Freiburg • Paris • London
New York • Bangalore • Bangkok
Singapore • Tokyo • Sydney

P.O. Box, CH–4009 Basel (Switzerland)
Printed in Switzerland on acid-free paper by
Reinhardt Druck, Basel
ISBN 978–3–8055–8311–4

KARGER

Fax +41 61 306 12 34
E-Mail karger@karger.ch
www.karger.com

Vol. 106, No. 2, 2007

nephron Journals

Contents

nephron Experimental Nephrology

KARGER

Fax +41 61 306 12 34
E-Mail karger@karger.ch
www.karger.com

Access to full text and tables of contents, including tentative ones for forthcoming issues: www.karger.com/nef_issues

nephron Physiology

nephron Clinical Practice

nephron
Experimental
Nephrology

Nephron Exp Nephrol 2007;106:e25–e26
DOI: 10.1159/000102014

Published online: June 6, 2007

Introduction

These have been exciting times for those of us interested in glomerular permeability. There has been a recent explosion of information concerning the structure and function of the mammalian glomerular capillary wall. In a striking example of the power of modern molecular medicine, detailed clinical studies on the diagnostic and prognostic importance of proteinuria have been augmented by molecular genetic studies in humans, gene targeting studies in rodents and sophisticated structural analyses 'in silico' to rapidly take forward our understanding of this key aspect of renal function. Translation of this new knowledge into diagnostic and therapeutic advances for patients is drawing near. Keeping up with this rapidly evolving literature challenges even those of us directly involved in the field, but is of interest to all involved in clinical and basic nephrology. Therefore, we believe that the time is right for a detailed review of current knowledge and for this reason we were delighted to be invited to bring together this supplement. We have assembled a veritable 'who's who' of experts in the field to bring you state of the art reviews.

The significance of proteinuria has been appreciated for generations, but the precise mechanisms whereby the mammalian kidney is able to filter enormous quantities of water and ions whilst not allowing the escape of any significant amounts of albumin have been poorly understood until recently. Anatomists had taught us that the glomerular capillary wall has three components: glomerular epithelial cells (podocytes), glomerular endothelial cells and between them the glomerular basement membrane (GBM). The GBM was shown to contain unique constituents and to possess marked negative electrostatic charge. Historically, the focus of studies on glomerular permeability, the normal 'sieve' action of the kidney and the disturbances associated with congenital or acquired diseases was for many years on the GBM. Alport syndrome and thin membrane nephropathy were considered developmental disorders primarily affecting the GBM, resulting in hematuria and sometimes also proteinuria. Acquired nephrotic syndrome, especially minimal change nephropathy, was thought to result from alterations of GBM charge.

Everything changed with the description in the late 1990s of single gene disorders primarily affecting podocytes which resulted in congenital nephrotic syndrome. A number of podocyte-specific genes were identified and the focus changed to the podocyte rather than the GBM. Rodent models have allowed analysis of the contribution of individual genes or regulatory elements, and exploitation of podocyte-specific gene promoters has allowed complex and elegant cell-specific studies. Selective injury to podocytes in rodents leads to a disease closely resembling focal segmental glomerulosclerosis (FSGS). Minimal-change nephropathy and FSGS have come to be considered as 'podocytopathies': this word is part of a whole new vocabulary including other terms such as podocytopenia, podocytology, podocyturia, etc., focusing attention on this unique, complex and aesthetically beautiful cell type. Podocytes are terminally differentiated cells with limited capacity for replication or repair: this may underlie their importance in progressive renal diseases and the prognostic significance of podocyte number in a

KARGER

Fax +41 61 306 12 34
E-Mail karger@karger.ch
www.karger.com

1660–2129/07/1062–0025$23.50/0

Accessible online at:
www.karger.com/nee

variety of disease situations. Drugs currently used in the treatment of glomerular disease have effects on podocytes; future approaches should harness these and aim to devise therapies capable of inducing podocyte repair or replacement.

Perhaps the focus on the podocyte has been too all-consuming: ironically, recent developments such as the work on Pierson syndrome have brought us full circle and reminded us that the GBM should not be forgotten. Furthermore, to this day the third component of the glomerular capillary wall, the glomerular endothelial cell, remains poorly understood and somewhat neglected. It is our view that the glomerular capillary wall is best viewed as a composite structure: the contribution of each of the three components may assume importance at various stages of development and in various disease states, but it is unwise to think of any of them in isolation.

Progressive chronic kidney disease is an increasingly important cause of morbidity and mortality in the developed world and a threat to the developing world. It places a major demand on health resources. Screening tests are based on detection of abnormalities in the urine, especially haematuria and proteinuria. It is sobering to realize how little we know about the structural and functional basis of those abnormalities. Our knowledge is improving fast, as explained in the articles that follow. Next we need to devise interventions that will allow prevention of progression: only then will the full value of the recent scientific advances be brought to fruition.

Peter W. Mathieson, Bristol, UK
Jeffrey H. Miner, St. Louis, Mo.

nephron
Experimental
Nephrology

Nephron Exp Nephrol 2007;106:e27–e31
DOI: 10.1159/000101789

Published online: June 6, 2007

Podocin Organizes Ion Channel-Lipid Supercomplexes: Implications for Mechanosensation at the Slit Diaphragm

Tobias B. Huber Bernhard Schermer Thomas Benzing

Renal Division, University Hospital Freiburg, Freiburg, Germany

Key Words
Slit diaphragm · Podocyte · Podocin · TRPC6 · Mechanosensor

Abstract
The slit diaphragm is part of a three-layered glomerular filter that prevents most proteins from entering the urinary space. As a highly specialized cell-cell contact the slit diaphragm has recently been appreciated as a signaling platform that regulates podocyte cell survival, polarity, endocytosis and cytoskeletal organization. Mutations in genes encoding slit diaphragm proteins contribute to human hereditary glomerular diseases. Recently, it was discovered that transient receptor potential ion channel *TRPC6* mutations cause familial glomerular disease. TRPC6 localizes to the slit diaphragm unexpectedly adding an ion channel to the list of signaling molecules functioning at this complex structure. Recent findings highlight a potential role for TRPC6 at the filtration barrier. TRPC6 has been identified as a sensor of mechanically and osmotically induced membrane stretch. In the context of the slit diaphragm signaling network, TRPC6 is clustered and regulated by a podocin-lipid complex that might translate mechanical tension to ion channel action. This review will summarize the most recent findings on protein-lipid supercomplexes at the slit diaphragm and discuss a potential novel function of the slit diaphragm as a mechanosensor.

Introduction

The podocyte is now recognized as the centerpiece of the glomerular filter. It spreads out across the capillary loops with regularly spaced, interdigitated foot processes that enwrap the glomerular capillaries. Interdigitating podocyte foot processes are bridged by a continuous membrane-like structure called the slit diaphragm [1].

In recent years several transmembrane, cytoskeletal and adaptor molecules have been identified as components of the complex molecular network of the slit diaphragm. Disruptions of this tightly regulated network result in loss of podocyte function and proteinuria. The study of human genetic glomerular diseases has helped to identify important players of the slit diaphragm such as nephrin, podocin, CD2AP, and TRPC6 [2]. Mutations affecting members of this protein complex result in nephrotic syndrome although the molecular basis for this is not clear and may differ in the different cases. Recently, mutations in TRPC6 were identified to cause focal segmental glomerulosclerosis (FSGS) with adult onset [3, 4]. All mutations exhibited a dominant mode of inheritance with variable penetrance. Subsequently, TRPC6 was localized in part to the insertion side of the slit diaphragm suggesting a potential involvement in slit diaphragm signaling pathways [4, 5].

KARGER

Fax +41 61 306 12 34
E-Mail karger@karger.ch
www.karger.com

1660–2129/07/1062–0027$23.50/0

Accessible online at:
www.karger.com/nee

Thomas Benzing
Renal Division, University Hospital Freiburg
Hugstetter Strasse 55, DE–79106 Freiburg (Germany)
Tel. +49 761 270 3559, Fax +49 761 270 3270
E-Mail thomas.benzing@uniklinik-freiburg.de

Mutations in podocin, a prohibitin homology (PHB)-domain protein, are the most common cause of hereditary nephrotic syndrome in humans [6]. The closest homologue of podocin is the PHB-domain protein MEC-2 in *Caenorhabditis elegans*. Interestingly, MEC-2 is part of a multiprotein channel complex that transduces sensation of gentle touch [7]. In touch receptor neurons, this channel complex is localized to regular puncta along the neuronal process and MEC-2 has been shown to regulate the MEC-4/MEC-10 ion channel complex in these puncta.

The recent identification of TRPC6 ion channels at the slit diaphragm in connection with the ability of podocin/Mec-2 proteins to regulate mechanosensitive ion channels leads to an intriguing novel concept of the slit diaphragm as a mechanosensor. This review will summarize the most recent evidence supporting the concept of mechanosensor functions of the slit diaphragm.

Components of a Mechanosensor: What Can We Learn from Worms?

The sense of touch relies on the ability of specialized sensory cells to convert mechanical stimulation into ionic currents. Mechanoreceptor cells respond to external force by opening ion channels. However, establishing the molecular identity of such channels has proved challenging. Traditional biochemical approaches are difficult to use because neither high-affinity ligands nor tissues rich in channels are available.

For several reasons, the nematode *C. elegans* is an ideal model in which to study the molecular machinery needed for electrical responses to touch: neurons that mediate touch avoidance are known and the mechanosensory function of the individual cells can be assayed in living animals. In addition, forward genetic screens allowed identification of genes required for the function of mechanosensation (MEC-1-MEC-12) [8]. The membrane protein products of four of these genes (MEC-2, MEC-4, MEC-6 and MEC-10) colocalize in touch neurons and can assemble voltage-independent, amiloride-sensitive Na^+ channels when coexpressed in *Xenopus laevis* oocytes [7]. MEC-4 and MEC-10 are DEG/ENaC proteins that form the pore of the channel complex. The podocin homologue MEC-2 associates with this complex to regulate its channel activity. In addition, the tubulin proteins MEC-7 and MEC-12 link the channel complex to the cytoskeleton [9]. This association appears essential for proper channel gating and provides a necessary anchoring point. The elegant genetic *C. elegans* studies have shown that specialized macromolecular complexes, encompassing mechanically gated ion channels, play a central role in the transformation of mechanical forces into a cellular signal. The genetic and molecular evidence collected in *C. elegans* has also highlighted the minimal components of the sensor apparatus that are required for conversion of a mechanical stimulus into a biological response: These are (1) channel proteins forming the pore of the channel complex, (2) associated molecules that physically interact with the channel subunits to shape and modulate their gating properties, and (3) cytoskeleton-associated molecules that bind to the channel complex and form a rigid intracellular grid that connects to the mechanotransducing ion channel, providing a necessary anchoring point.

Translating these findings to the slit diaphragm, it becomes obvious that the macromolecular protein complex of the slit diaphragm contains all components needed to form a mechanosensor apparatus: (1) TRPC6 ion channels as mechanosensitive ion channels, (2) the MEC-2 homologue, podocin, to transduce mechanical force to the channel and modulate gating properties, and (3) a very tight cytoskeletal anchoring of the slit diaphragm to provide the necessary anchoring point.

In the following sections, we will discuss the evidence that these components might indeed act together to form a functional mechanosensory apparatus.

TRPC Channels Are Mechanosensitive Channels

To date there are at least two distinct classes of mechanically gated ion channels known: the degenerin (DEG)/epithelial Na^+ channel (ENaC) family and the transient receptor potential (TRP) family of ion channels. TRPC6 is a member of the large superfamily of TRP cation channels that transduce a remarkable spectrum of signals including temperature, osmolarity and touch [10]. The closely related subgroup comprising TRPC3, TRPC6, and TRPC7 channels are directly activated by diacylglycerol through a PKC-independent mechanism. TRPC channels consist of amino-terminal ankyrin repeats and potential coiled-coil domains followed by six transmembrane domains, and a short TRP box and additional coiled-coil domains at the carboxy-terminus. All known FSGS-associated TRPC6 mutations are located on the amino- or carboxy-terminal cytoplasmic parts of the protein. When expressed in different in vitro systems these mutations cause either gain of function or comparable current amplitudes to that of wild-type channels [3, nephritis, crescentic glomerulonephritis and minimal change disease' all refer to distinguishable morphological changes in the glomerular architecture. In marked contrast to these detailed morphological descriptions, we still know very little about the molecular composition of the glomerulus and how it changes during disease. However, this situation is now beginning to change. Transcript profiling technologies have been improved to the extent that they are both genome-wide and applicable to minute amounts of biological material, such as single or few glomeruli. During the past few years various methodologies have been applied to map the glomerular transcriptome, and some proteomic efforts have also been made. These studies have already yielded new insights into the normal molecular composition of the glomerulus and its various structures, such as the podocyte foot processes and slit diaphragms, but there is still some way to go to complete the map of the glomerular transcriptome and proteome. Furthermore, huge additional efforts are required to assign functions to the many new glomerular proteins discovered. However, en route, glomerular molecular profiling of various experimental conditions of impaired glomerular function in animals, and of various human primary or secondary glomerular diseases, will aid in the classification of glomerular diseases. It will also help to reveal pathogenic pathways and cause-effect relationships in these pathways, thereby providing a basis for future rational drug design. Finally, glomerular molecular profiling may be useful for the stratification of patient populations in clinical trials, and for monitoring glomerular responses to therapy.

The Glomerular Transcriptome: A Tiny Fraction of the Whole Kidney Transcriptome

Whole-kidney transcript profiles have been determined and found useful in the characterization of pathological responses in the kidney following acute allograft rejection [1]. Whole-kidney transcript profiling has also been applied to renal conditions that injure the glomeruli, such as diabetes [2], but the importance of these recorded changes in relation to the pathogenic events in diabetic nephropathy remains to be established. Glomeruli make up less than 10% of the kidney volume, and the glomerular transcriptome constitutes only a minor proportion of the total kidney transcriptome. Most likely the whole-kidney profiles do not provide sufficient sensitivity and degree of resolution to allow determination of the glomerular transcriptome. Indeed, attempts to measure the expression of podocyte-specific transcripts using microarray analysis or EST profiling of whole kidney failed for all glomerular markers except podocin, which is the single most highly expressed marker for podocytes [3, 4].

Methods to Determine the Glomerular Transcriptome

Determination of the glomerular transcriptome requires pure isolates of glomeruli (fig. 1). Glomerulus isolation is classically accomplished by sieving through appropriately sized meshes, or by micro-dissection. To facilitate isolation of glomeruli from mice at large scale and to high degree of purity, we developed a method involving magnetic bead perfusion followed by tissue digestion and glomerulus isolation by a magnet [5]. This method works well in postnatal mice but also in embryos, allowing for the isolation of highly pure preparations of glomeruli of different developmental stages [4, 6]. The efficiency of the method, allowing the collection of virtually all glomeruli of a mouse (≈20,000 from adults and a few hundred from late embryos) in a single pure isolate, has significantly increased the possibilities to derive sufficient quantities of high-quality glomerular RNA for, for example, transcription profiling.

Glomerular transcript profiles have been derived from both human and mouse using different profiling technologies (fig. 1). Human glomerular profiles have been derived using serial analysis of gene expression (SAGE) of micro-dissected nephrons [7], and by DNA microarray analysis of glomeruli isolated by sieving [8, 9] or micro-dissection [10], including laser capture [11, 12]. Mouse glomerular profiles have been determined using DNA microarray analysis of glomeruli isolated by magnetic beads [4, 6] or sieving [13], as well as by comparative EST profiling of glomerular and whole-kidney cDNA libraries [3].

Glomerular Transcriptional Changes in Developmental Disorders and Disease

Glomerular transcription profiling by global technologies offers the possibility of elucidating downstream transcriptional targets of signaling pathways and transcription factors in defined cell populations within the glomerulus. Moreover, secondary changes to a primary insult in, for example, podocytes may be mapped in glo-

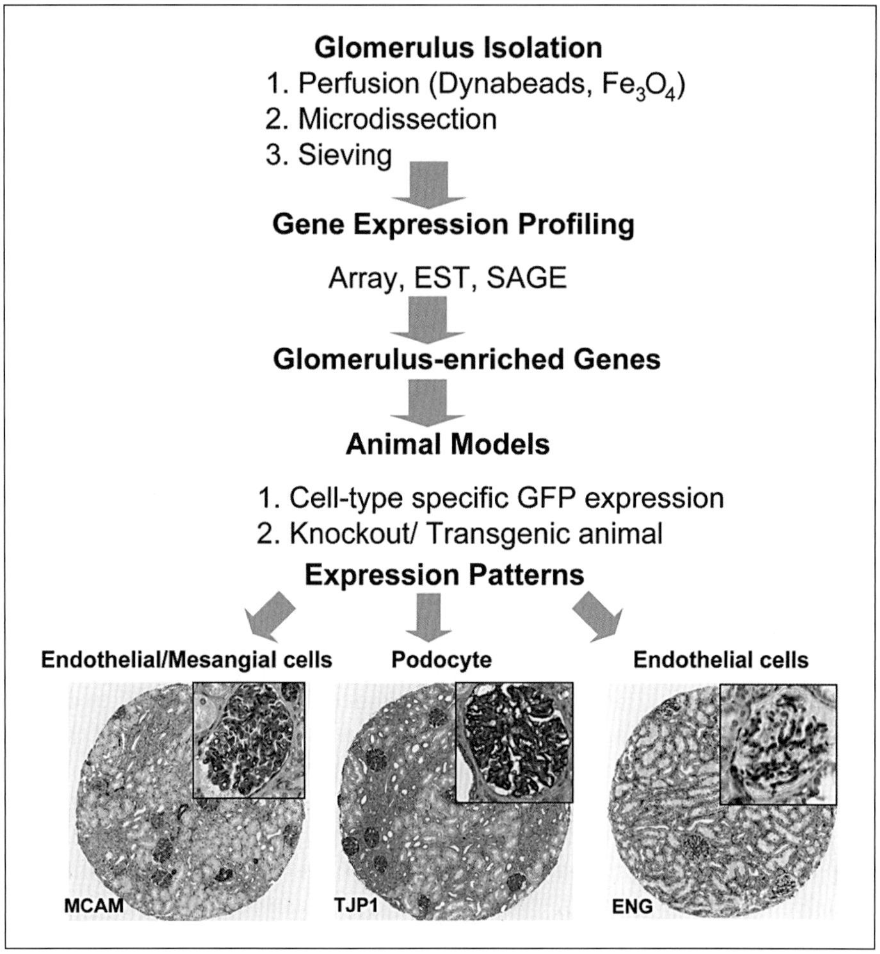

Fig. 1. Flow chart of glomerular profiling. Different methodologies for glomerulus isolation and gene expression profiling are indicated. Combined with selected animal models representing specific gene and cell ablations, or green fluorescent protein (GFP) marker expression, genes and proteins may be assigned to different glomerular cell types. The examples of histological images represent the antibody-based proteomic effort Human Proteome Resource (HPR) and are taken from the publicly accessible Human Protein Atlas (www.proteinatlas.org).

merular endothelial and mesangial cells. Thus, the comparative transcription profiling of numerous defined genetic mutants (e.g. knockout mice) representing genes with an assumed regulatory role in the glomerulus holds the promise of revealing epistatic and cause-effect relationships. Microarray analyses of glomerular isolates from knockout embryos of the podocyte transcription factors, Pod1 and Foxc2, provide two such examples [4, 6]. In these studies, several podocyte-specific genes could be identified as targets (direct or indirect) for both Pod1 and Foxc2. However, direct comparison of the data was not possible due to the use of different microarray platforms in the two studies. Additional profiling of glomeruli from these and other mutants of podocyte transcription factors (such as Lmx1b and Mafb) using common technical platforms will likely reveal hierarchical relationships in podocyte transcriptional regulation. Thus, while recently published data holds promise, it is important to note that the possibilities of cross-comparing profiles from different mutants and conditions will heavily rely on the utilization of common techniques for tissue sampling (e.g. glomerulus isolation by magnetic beads) and transcription profiling (e.g. affymetrix arrays).

A few glomerular diseases or disease-like conditions have been profiled up to date, but similar to the above mentioned studies of Pod1 and Foxc2 knockouts, differences in glomerulus isolation and profiling platforms make direct comparisons of the results impossible. Human studies conducted thus far report transcript profiles from glomeruli in diabetes [8], lupus nephritis [11] and obesity-related glomerulopathy [10]. Mouse studies include various models of diabetes [12–14]. Rat studies include anti-Thy1 nephritis [15]. Although various differences were revealed between healthy and diseased glomeruli in these studies, they currently remain purely descriptive and they do not pinpoint common pathogenic pathways or responses to disease states.

Novel Glomerular Cell Type-Specific Markers

Even if profiling of pure glomeruli is a huge improvement over whole-kidney profiling, the presence of three distinct cell types in the glomerulus complicates the interpretations. Ideally one would like to be able to assign the transcriptional changes observed in a disease or experimental glomerular injury to individual glomerular cell types. The recent identification of several mutations underlying monogenic nephrotic syndromes show that the genes affected are all expressed specifically by podocytes [16]. Moreover, several of the encoded proteins can be directly linked with the formation and function of the podocyte slit diaphragm. In marked contrast, information is lacking about the molecular etiology of the glomerular complications resulting from diabetes, high blood pressure, Ig complex formation and other systemic aberrations. Given the critical role of podocytes in glomerular function, and the identification of disease-causing mutations in podocyte-specific genes, it is tempting to speculate that podocytes are also the targets of systemic insults (e.g. hyperglycemia or advanced glycation end products in diabetes). However, we cannot exclude an involvement of the glomerular endothelial cells or mesangial cells. In fact, mouse genetics have demonstrated that proteinuria, glomerulosclerosis and ESRD can result from a primary insult to endothelial or mesangial cells [17–19].

Our own analysis using spotted microarrays containing cDNA clones representing some 6,000 different glomerulus-expressed genes have revealed the existence of more than 300 different genes that were significantly overexpressed in glomeruli in comparison with nonglomerular kidney tissue [4]. As a first step towards assigning these to the different glomerular cell types, we separated genetically green fluorescent protein-labeled podocytes from the rest of the glomerulus (endothelium and mesangium) and generated profiles from the two cell pools (fig. 1). This led to the identification of 49 transcripts with high expression in the podocytes and 18 transcripts with high expression in the rest of the glomerulus [4]. Interestingly, most of the known podocyte markers were included among the 49 and several known markers for mesangial cells were included among the 18. In situ hybridization for some of the novel podocyte-specific marker candidates confirmed podocyte-specific expression for five of them and mesangial-specific expression for another three [4]. In a follow-up study, we compared expressed sequence tag (EST) abundances between whole-kidney and glomerular cDNA libraries to identify another six glomerular markers, of which three could be assigned to podocytes and three to the endothelial/mesangial compartment. Additional studies of this kind will most likely increase further the lists of genes that are specifically expressed by individual glomerular cell types, thereby helping the interpretation of changes in the whole-glomerulus profiles.

Analysis of the Glomerular Proteome by Large-Scale Antibody Approach

The Human Protein Atlas (HPA) is a database of human protein expression patterns in normal organs and various solid tumors [20]. The expression patterns have been determined by immunohistochemistry and can be accessed online (www.proteinatlas.org). The current version (2.0) contains data for more than 1,500 antibodies, representing some 1,000 different gene products. Searching the Protein Atlas for examples where the glomerular expression of a certain human protein is higher than in surrounding renal tissues reveals 176 proteins (ref. [3] and www.proteinatlas.org; see fig. 1 for examples). With future updates of the Atlas, this list of glomerular protein markers is likely to expand significantly.

Mono-specific antibodies provide means of determining the cellular localization of novel glomerular proteins. Moreover, their subcellular distribution may be determined using immuno-electron microscopy. This is important because a discrete subcellular localization, for example to the podocyte foot processes or slit diaphragm, may suggest a role in the development or maintenance of the kidney filter. We have begun generating antibodies against a catalog of >500 proteins whose transcripts are enriched in the glomerulus. We recently published the analysis of five such antibodies, which revealed the expression and subcellular distribution of the novel glomerular proteins dendrin, Ehd3, Sh2d4a, Plekhh2 and 2310066E14Rik [21]. This led to a series of striking observations. First Ehd3 was confirmed as the first glomerular endothelium-specific protein, i.e. it is expressed in glomerular endothelium but not at detectable levels in other kidney endothelial cells. Dendrin, Sh2d4a, Plekhh2 and 2310066E14Rik were all expressed by podocytes, and dendrin, Sh2d4a and Plekhh2 were further sub-localized to the slit diaphragm, suggesting a role for these proteins in size-selective filtration.

Outlook

Days are still early for glomerular molecular profiling, but current efforts in the areas of transcriptome and proteome analysis show great promise, and most likely our knowledge about the gene and protein expression complexity in the glomerulus will grow steadily over the coming years. This will lead to new fundamental knowledge about the structure and development of the renal filter of its cellular and extracellular components. Moreover, comprehensive profiling of various models of disturbed glomerular development and function, as well as human diseases affecting glomerular function, and of animal models of these diseases, will lead to the identification of critical developmental and maintenance processes, and unique and common pathogenic pathways involved in glomerular disease. Some of these pathways will hopefully lend themselves to targeting by pharmacological compounds.

Acknowledgements

Own work referenced herein was kindly supported by grants from the Knut and Alice Wallenberg, Novo Nordisk, Söderberg, Inga-Britt and Arne Lundberg Strategic Research, Juselius and Hedlund Foundations, the EU 6th framework integrated project Lymphangiogenomics, LSHG-CT-2004-503573, and the Swedish Science Council.

References

1 Sarwal M, Chua MS, Kambham N, Hsieh SC, Satterwhite T, Masek M, Salvatierra OJ: Molecular heterogeneity in acute allograft rejection identified by DNA microarray profiling. N Engl J Med 2003;349:125–138.

2 Wada J, Makino H, Kanwar YS: Gene expression and identification of gene therapy targets in diabetic nephropathy. Kidney Int 2002;61:73–78.

3 He L, Sun Y, Patrakka J, Mostad P, Norlin J, Xiao Z, Andrae J, Tryggvason K, Samuelsson T, Betsholtz C, Takemoto M: Glomerulus-specific mRNA transcripts and proteins identified through kidney expressed sequence tag database analysis. Kidney Int 2007; in press.

4 Takemoto M, He L, Norlin J, Patrakka J, Xiao Z, Petrova TV, Bondjers CJA, Wallgard E, Sun Y, Samuelsson T, Mostad P, Lundin S, Miura N, Sado Y, Alitalo K, Quaggin SE, Tryggvason K, Betsholtz C: Large-scale identification of genes implicated in kidney glomerulus development and function. EMBO J 2006;25:1160–1174.

5 Takemoto M, Asker N, Gerhardt H, Lundkvist A, Johansson BR, Saito Y, Betsholtz C: A new method for large scale isolation of kidney glomeruli from mice. Am J Pathol 2002; 61:799–805.

6 Cui S, Li C, Ema M, Weinstein J, Quaggin SE: Rapid isolation of glomeruli coupled with gene expression profiling identifies downstream targets in Pod1 knockout mice. J Am Soc Nephrol 2005;6:1587–1597.

7 Chabardes-Garonne D, Mejean A, Aude J-C, Cheval L, Di Stefano A, Gaillard M-C, Imbert-Teboul M, Wittner M, Balian C, Anthouard V, Robert C, Segurens B, Wincker P, Weissenbach J, Doucet A, Elalouf JM: A panoramic view of gene expression in the human kidney. Proc Natl Acad Sci USA 2003;100: 13710–13715.

8 Baelde HJ, Eikmans M, Doran PP, Lappin DWP, de Heer E, Bruijn JA: Gene expression profiling in glomeruli from human kidneys with diabetic nephropathy. Am J Kidney Dis 2004;43:636–650.

9 Higgins JPT, Wang L, Kambham N, Montgomery K, Mason V, Vogelmann SU, Lemley KV, Brown PO, Brooks JD, van de Rijn M: Gene expression in the normal adult human kidney assessed by complementary DNA microarray. Mol Biol Cell 2004;15:649–656.

10 Wu Y, Liu ZJ, Xiang Z, Zeng C, Chen Z, Ma X, Li L: Obesity-related glomerulopathy: insights from gene expression profiles of the glomeruli derived from renal biopsy samples. Endocrinology 2006;147:44–50.

11 Peterson KS, Huang J-F, Zhu J, D'Agati V, Liu X, Miller N, Erlander MG, Jackson MR, Winchester RJ: Characterization of heterogeneity in the molecular pathogenesis of lupus nephritis from transcriptional profiles of laser captured glomeruli. J Clin Invest 2004;113:1722–1733.

12 Naito Y, Uchiyama K, Mizushima K, Kuroda M, Akagiri S, Takagi T, Handa O, Kokura S, Yoshida N, Ichikawa H, Takahashi J, Yoshikawa T: Microarray profiling of gene expression patterns in glomerular cells of astaxanthin-treated diabetic mice: a nutrigenomic approach. Int J Mol Med 2006;18:685–695.

13 Makino H, Miyamoto Y, Sawai K, Mori K, Mukoyama M, Nakao K, Yoshimasa Y, Suga S: Altered gene expression related to glomerulosclerosis and podocyte structure in early diabetic nephropathy of db/db mice and its restoration by pioglitazone. Diabetes 2006; 55:2747–2756.

14 Susztak K, Böttinger E, Novetsky A, Liang D, Zhu Y, Ciccone E, Wu D, Dunn S, McCue P, Sharma K: Molecular profiling of diabetic mouse kidney reveals novel genes linked to glomerular disease. Diabetes 2004;53:784–794.

15 Sadlier DM, Ouyang X, McMahon B, Mu W, Ohashi R, Rodgers K, Murray D, Nagagawa T, Godson C, Doran P, Brady HR, Johnson RJ: Microarray and bioinformatic detection of novel and established genes expressed in experimental anti-Thy1 nephritis. Kidney Int 2005;68:2542–2561.

16 Tryggvason K, Patrakka J, Wartiovaara J: Hereditary proteinuria syndromes and mechanisms of proteinuria. N Engl J Med 2006;354:1387–1401.

17 Eremina V, Sood M, Haigh J, Nagy A, Lajoie G, Ferrara N, Gerber HP, Kikkawa Y, Miner JH, Quaggin SE: Glomerular-specific alterations of VEGF-A expression lead to distinct congenital and acquired renal diseases. J Clin Invest 2003;111:707–716.

18 Lindblom P, Gerhardt H, Liebner S, Abramsson A, Enge M, Hellstrom M, Backstrom G, Fredriksson S, Landegren U, Nystrom HC, Bergstrom G, Dejana E, Ostman A, Lindahl P, Betsholtz C: Endothelial PDGF-B retention is required for proper investment of pericytes in the microvessel wall. Genes Dev 2003;17:1835–1840.

19 Bjarnegård M, Enge M, Norlin J, Gustafsdottir SM, Fredriksson S, Abramsson A, Takemoto M, Gustafsson E, Fässler R, Betsholtz C: Endothelium-specific ablation of PDGFB leads to pericyte loss and glomerular, cardiac and placental abnormalities. Development 2004;131:1847–1857.

20 Uhlen M, Bjorling E, Agaton C, et al: A human protein atlas for normal and cancer tissues based on antibody proteomics. Mol Cell Proteomics 2005;4:1920–1932.

21 Patrakka J, Xiao Z, Nukui M, Takemoto M, He L, Oddsson A, Perisic L, Kaukinen A, Al-Khalili Szigyarto C, Uhlén M, Jalanko H, Betshotz C, Tryggvason K: Expression and subcellular distribution of novel glomerulus-associated proteins dendrin, Ehd3, Sh2d4a, Plekhh2 and 2310066E14Rik. J Am Soc Nephrol 2007; e-pub ahead of print.

nephron
Experimental
Nephrology

Nephron Exp Nephrol 2007;106:e37–e43
DOI: 10.1159/000101791

Published online: June 6, 2007

Autoepitopes and Alloepitopes of Type IV Collagen: Role in the Molecular Pathogenesis of Anti-GBM Antibody Glomerulonephritis

Dorin-Bogdan Borza

Division of Nephrology, Department of Medicine, Vanderbilt University School of Medicine, Nashville, Tenn., USA

Key Words
Autoantibody · Alloantibody · Goodpasture disease · Anti-glomerular basement membrane disease epitope mapping

Abstract
Anti-glomerular basement membrane (anti-GBM) antibodies elicited by autoimmune or alloimmune mechanisms are associated with aggressive forms of rapid progressive glomerulonephritis. Pathogenic anti-GBM autoantibodies and alloantibodies target the noncollagenous (NC1) domains of the α3α4α5(IV) collagen, a major GBM component. In autoimmune anti-GBM glomerulonephritis, a breakdown of immune self-tolerance leads to the activation of autoreactive B and T cells recognizing epitopes within the α3NC1 subunit. In the GBM, the conformational epitopes targeted by anti-GBM autoantibodies are structurally sequestered within the α3α4α5NC1 hexamer complex formed upon assembly of collagen IV chains into trimeric molecules and networks. Autoantibodies selectively bind to and dissociate a subset of α3α4α5NC1 hexamers composed of monomer subunits, whereas hexamers containing NC1 dimer subunits are resistant to dissociation by autoantibodies. The crypticity of α3NC1 autoepitopes suggests that self-tolerance to α3(IV) collagen is broken by structural alterations of the native α3α4α5NC1 hexamer that unmask normally sequestered epitopes, triggering an autoimmune reaction. Post-transplant anti-GBM nephritis in the renal allograft of transplanted Alport patients is mediated by an alloimmune reaction to the NC1 domains of α3α4α5(IV) collagen, present in the allograft GBM but absent from Alport basement membranes. Alloantibodies from patients with autosomal-recessive Alport syndrome predominantly bind to the α3NC1 domain, whereas alloantibodies from X-linked Alport patients target preferentially, though not exclusively, epitopes within the α5NC1 subunit. The accessibility of the alloantigenic sites within the α3α4α5NC1 hexamers, contrasting with the crypticity of autoantigenic sites, suggest that different molecular forms of α3α4α5(IV) collagen initiate the immunopathogenic responses in the two forms of anti-GBM disease. Advances in elucidating the structure of the GBM antigen and the identification of the pathogenic B and T cell epitopes, along with new insights into the pathogenic mechanisms at cellular and molecular level will facilitate the development of targeted strategies for prevention, detection, and treatment of human anti-GBM antibody glomerulonephritis.

KARGER

Fax +41 61 306 12 34
E-Mail karger@karger.ch
www.karger.com

Accessible online at:
www.karger.com/nee

Dorin-Bogdan Borza, PhD
S-3223 Medical Center North
Vanderbilt University Medical School
Nashville, TN 37232-2372 (USA)
Tel. +1 615 322 2089, Fax +1 615 343 7156, E-Mail Dorin-Bogdan.Borza@vanderbilt.edu

Introduction

Glomerulonephritis is mediated by an abnormal engagement of adaptive and innate immune responses within the glomerulus. The glomerular basement membrane (GBM), an essential component of the blood ultrafiltration barrier, is a prominent target of immune-mediated glomerular injury in human disease and animal models. Anti-GBM antibodies circulating in patient plasma and deposited in a characteristic linear pattern along the basement membranes of glomerular capillary loops are hallmark features in anti-GBM antibody disease, one of the most aggressive forms of glomerulonephritis. The fulminant course of this disease makes early diagnosis and prompt treatment by plasma exchange and immunosuppressants (corticosteroids and cyclophosphamide) critical for preserving patients' glomerular function. The high morbidity and mortality (over 50% lethality in untreated patients) associated with anti-GBM antibody nephritis underscores the clinical importance of this condition.

Anti-GBM antibody nephritis has emerged as a paradigm for immune-mediated glomerular injury because of the detailed knowledge of the antigen structure and the comprehensive characterization of the pathogenic epitopes, which – together with the availability of good animal models – provide a unique framework for investigating the molecular mechanisms of disease. Pathogenesis of anti-GBM antibody nephritis involves either autoimmune mechanisms, in the spontaneous form of disease, or alloimmune mechanisms, in the form of the disease that may occur in renal transplants in patients whose primary renal disease was Alport syndrome. Both autoantibodies and alloantibodies specifically target epitopes residing within the noncollagenous (NC1) domains of a subset of collagen IV chains present in the GBM. However, the recent characterization of the anti-GBM autoepitopes and alloepitopes revealed differences in their location in the supramolecular structure of the antigen and accessibility in tissues, indicating that distinct upstream events initiate the immunopathogenic processes in the two forms of disease. After binding to their targets in the GBM, the antibodies (typically IgG) initiate local tissue injury by activating the complement system and by recruiting effector immune cells, such as macrophages and neutrophils, via immunoglobulin Fc receptors. As common downstream mediators of inflammation are implicated in autoimmune and alloimmune anti-GBM disease, the clinical presentation (rapid progressive glomerulonephritis) and pathology findings (crescentic and necrotizing glomerulonephritis) are similar.

Autoimmune Anti-GBM Antibody Disease

Autoimmune anti-GBM antibody disease, including Goodpasture disease (see below), occurs spontaneously with an incidence of 0.5–1 case per million per year and is usually linked to a breakdown of immune self-tolerance to the α3(IV) collagen chain. The majority of patients have circulating IgG autoantibodies that bind relatively specifically and with high affinity to the NC1 domain of α3(IV) collagen. The α3(IV) chain has a tissue-restricted distribution, and is accompanied by α4 and α5(IV) chains. The three chains co-assemble into an extensively cross-linked supramolecular network that is the native form of autoantigen in tissues. The α3α4α5(IV) collagen network is a major structural component of the GBM, and is also found in the basement membranes of lung alveoli, seminiferous tubules, lens capsule, and choroid plexus.

Anatomical barriers that prevent the access of anti-GBM autoantibodies to autoantigen restrict the sites of autoimmune injury. Circulating autoantibodies bind most readily to α3(IV) collagen in the GBM. This site is particularly vulnerable because glomerular capillaries are specialized for ultrafiltration of large volumes of blood, and it is proposed that plasma proteins of IgG size can permeate freely across glomerular endothelium, presumably via the fenestrations. In other capillary beds, tight junctions of endothelial cells comprise an anatomical barrier that prevents circulating autoantibodies from reaching autoantigen in the underlying basement membranes. Between half and two thirds of patients with anti-α3NC1 autoantibodies have Goodpasture (GP) disease, a combination of glomerulonephritis and lung hemorrhage. Pulmonary hemorrhage is mediated by anti-α3NC1 autoantibodies binding to antigen in alveolar basement membranes. Although lung capillaries are normally impervious to large molecules such as IgG, certain pathophysiological conditions increase their permeability. In animal models, binding of anti-GBM antibodies to alveolar basement membranes is promoted by experimental treatments such as oxygen toxicity, pro-inflammatory cytokines, or intratracheal instillation of gasoline. In human GP disease, lung injury may be triggered by inhaled environmental toxins, such as cigarette smoking.

Molecular barriers restricting the access of anti-GBM/GP autoantibodies to target epitopes emerge from the assembly of α3(IV) collagen into a quaternary structure. The NC1 domains direct the specific assembly of collagen IV chains into heterotrimeric molecules and mediate the end-to-end association of two heterotrimers, forming a stable NC1 hexamer complex. Consequently, the epitopes recognized by anti-GBM/GP autoantibodies become structurally sequestered (cryptic) within the α3α4α5NC1 hexamers. Experimentally, the crypticity of the autoepitopes manifests as a several-fold increase in autoantibody binding to α3NC1 subunits compared to native α3α4α5NC1 hexamers. Three converging lines of investigation provided insights into the molecular mechanisms sequestering the autoepitopes: identification of α3NC1 regions recognized by autoantibodies [1, 2], knowledge of the organization of subunits within the α3α4α5NC1 hexamers [3], and determination of the X-ray crystal structure of homologous α1α2NC1 hexamers [4, 5].

Anti-GBM/GP autoantibodies have restricted specificity among patients, targeting a small number of conformational α3NC1 epitopes, which have been identified by using recombinant α1/α3 chimeric NC1 domains expressed in mammalian cells [1, 2]. The immunodominant epitope, E_A, is encompassed by α3NC1 residues 17–31. A homologous region at α3NC1 residues 127–141 encompasses the E_B epitope, recognized by a distinct subpopulation of autoantibodies present only in some patient sera (fig. 1a). The nephritogenic potential of these epitopes has been experimentally verified in WKY rats, which developed autoimmune glomerulonephritis after immunization with chimeric α1/α3NC1 domains encompassing the E_A or E_B regions [6]. The crypticity of the E_A and E_B autoepitopes is due to their location within the quaternary structure of the α3α4α5NC1 hexamer, partly hidden along ridges adjacent to α5NC1 and α4NC1 subunits, respectively [3]. Within the E_A region, four hydrophobic residues, Ala^{18}, Ile^{19}, Val^{27} and Pro^{28}, are required for autoantibody binding [7]. Amino acids in the E_A region together with Gln^{57} form another autoepitope [8]. Key autoepitope residues, such as Val^{27} and Gln^{57}, become partially buried in the NC1 hexamer, which may sterically hinder autoantibody binding [3].

An investigation of the biochemical mechanisms by which autoantibodies can bind to cryptic epitopes has identified two distinct subsets of α3α4α5NC1 hexamers in the human GBM [9]. Under native conditions, anti-GBM/GP autoantibodies selectively react with M-α3α4α5 NC1 hexamers – composed of α3NC1, α4NC1 and α5NC1 monomer subunits only – with concomitant dissociation of the quaternary structure (fig. 1b). In contrast, D-α3α4α5 NC1 hexamers, containing α3/α5NC1 and α4/α4NC1 dimers in addition to monomer subunits, do not bind autoantibodies unless pre-dissociated by low pH or protein denaturants. The inertness of D-α3α4α5 hexamers under native conditions indicates that cross-links within the NC1 dimer subunits reinforce the quaternary structure of the antigen, averting dissociation by autoantibodies. The molecular identity of the crosslink has been established for the homologous α1α2NC1 hexamers, in which Met^{93} is covalently bonded to Lys^{211} from an adjoining NC1 domain [5], post-translationally modified to hydroxylysine [10]. These residues are conserved in all six mammalian NC1 domains, and homologous cross-links connect α3/α5NC1 and α4/α4NC1 dimers within D-α3α4α5NC1 hexamers [Vanacore et al., manuscript in preparation]. Because circulating anti-GBM autoantibodies do not bind to native M-hexamers or D-hexamers, neither isoform appears to be the proximal antigen initiating disease. Instead, autoimmunity is likely triggered by pathogenic conditions such as reactive oxygen species [11] that promote the exposure of cryptic α3NC1 epitopes and aberrantly engage the adaptive immune system. The resulting autoantibodies can subsequently bind to M-α3α4α5 NC1 hexamers in the human GBM, though not to the more abundant D-α3α4α5 hexamers.

Although circulating and kidney-bound autoantibodies are the most prominent feature of anti-GBM disease, there is increasing evidence that sensitized autoreactive T cells are also essential pathogenic effectors. A role for CD4+ helper T cells, presumably in the initial breaking of tolerance at the T cell level that is thought to underlie autoimmunity, has long been suspected based on strong association between anti-GBM disease and HLA class II alleles, such as DRB1*1501. Expression of α3(IV) collagen in human thymus suggests that T cell tolerance is centrally established by deletion of autoreactive cells [12]. However, patients with acute anti-GBM/GP disease have an increased frequency of CD4+ T cells reactive with α3NC1, which normalizes during remission [13]. Patients' lymphocytes proliferate in response to several α3NC1 peptides, most commonly $\alpha3_{71-90}$ and $\alpha3_{131-150}$, and secrete interferon-γ, a Th1 cytokine, but not IL-4, a Th2 cytokine [14]. In mouse models, the susceptibility of various strains to autoimmune glomerulonephritis is associated with the emergence of a Th1-type polarization [15].

Activated Th cells have multiple effector roles. The profile of cytokines secreted by polarized Th cells influence the immunoglobulin class switch in B cells. Th1-type cytokines stimulate production of IgG subclasses (IgG1 in humans, IgG2a in mice) that bind complement and have high affinity for Fc receptors. In human disease, isolated case reports suggest that IgG1 but not IgG4 (a Th2-type subclass) autoantibodies are pathogenic. In WKY rats, glomerulonephritis can be passively induced by anti-GBM monoclonal antibodies of Th1-type isotypes IgG2a and IgG2b, but not by IgG1, a Th2-type subclass [16]. Besides modulating the humoral immune responses, CD4+ T lymphocytes are found in increased numbers in kidneys with glomerulonephritis and may be directly pathogenic. The nephritogenic potential of autoreactive CD4+ T cells has been demonstrated in WKY rats immunized with misfolded α3NC1 or with a nephritogenic peptide containing a T cell epitope (α3NC1 residues 28–40), which elicited T cell-mediated glomerulonephritis without IgG deposition along the GBM [17, 18].

T cells from convalescent patients have a strikingly different phenotype, secreting the regulatory cytokine IL-10 upon stimulation with α3NC1 peptides [14]. A distinct subset of CD4+CD25+ regulatory T cells (Treg) emerges that is reactive to α3NC1 and can modulate T cell responses to antigen [19]. The reduced frequency of autoreactive T cells and the emergence of a new population of regulatory T cells in treated patients suggest that tolerance to α3NC1 can be effectively reestablished following acute disease, which may explain the rarity of relapse. Targeted stimulation and expansion of the pool of regulatory T cells may provide the basis for new, rationally designed, specific therapies in patients with autoimmune anti-GBM disease.

Alloimmune Anti-GBM Antibody Nephritis

The alloimmune form of anti-GBM disease occurs only in some patients with Alport syndrome after a kidney transplant. The patients at risk have defects in COL4A3, COL4A4 or COL4A5 genes that completely prevent the normal assembly of the α3α4α5(IV) collagen in their basement membranes. Defects may be nonsense mutations, out-of-frame insertions or deletions, or large gene rearrangements. In these patients, the absence of α3α4α5(IV) collagen from the GBM is provisionally compensated by persistence of the embryonic α1α2(IV) network, but the abnormal composition of the Alport GBM leads to a gradual deterioration of its ultrastructure, compromising the integrity of the glomerular filtration barrier. Consequently, patients develop hematuria, proteinuria, and slowly progress to end-stage renal failure, requiring dialysis or transplantation. Among Alport patients receiving a kidney transplant, ~3–5% develop post-transplant anti-GBM alloantibody nephritis, due to a severe alloimmune reaction to 'foreign' α3α4α5(IV) collagen present only in the allograft GBM. Most affected patients lose allograft function and have a very high risk of disease recurrence in subsequent transplants.

Alport post-transplant alloantibodies exhibit a broad spectrum of specificities for various NC1 subunits of α3α4α5(IV) collagen. A key determinant of the alloantibody specificity is the nature of the primary molecular defect causing Alport syndrome. In patients with autosomal-recessive Alport syndrome (ARAS), caused by homozygous mutations in COL4A3 or COL4A4 genes, circulating anti-GBM alloantibodies invariably react with α3NC1 [20, 21], and pathogenic allograft-bound alloantibodies react with both α3NC1 and α4NC1 [22]. Notably, α5NC1 is not a target of ARAS alloantibodies, presumably because immune tolerance is established toward the α5(IV) collagen chain, which continues to assemble normally with the α6(IV) chain. An analysis of the epitope specificity of kidney-eluted ARAS alloantibodies has identified one alloepitope jointly composed of the E_A and E_B regions of α3NC1 [22]. Parts of these regions that are proximal to one another are distant from, and hence not shielded by, adjacent NC1 domains (dotted circle in fig. 1a). Thus, the ARAS alloepitope appears to be accessible in the α3α4α5NC1 hexamer, in contrast to the cryptic α3NC1 autoepitopes separately encoded by the E_A and E_B regions. This implies that in ARAS patients, not only T cells but also B cells specific for accessible α3NC1 epitopes fail to undergo deletion, anergy or receptor editing, and play an active role in the pathogenesis of ARAS post-transplant nephritis, but not in autoimmune anti-GBM disease.

In patients with X-linked Alport syndrome (XLAS), caused by defects in the COL4A5 gene, the common target of all anti-GBM alloantibodies is the α5NC1 domain [20, 21]. COL4A5 mutations cause the loss of α5(IV) chain and prevent the assembly of α3α4α5(IV) and α5α6(IV) networks, so that effective tolerance to α5NC1 epitopes cannot be established in either the B or T cell compartments. Some, though not all XLAS patient sera additionally contain alloantibodies to α3NC1 [20, 21, 23], possibly generated secondarily by epitope spreading. Pathogenic XLAS alloantibodies eluted from kidney allografts bind almost exclusively to conformational epitopes within α5NC1 [23]. Two immunodominant alloepitopes have been identified within α5NC1 residues 1–45 and 114–168 (fig. 1a). These alloepitopes are also the major targets of circulating XLAS alloantibodies. Alloanti-

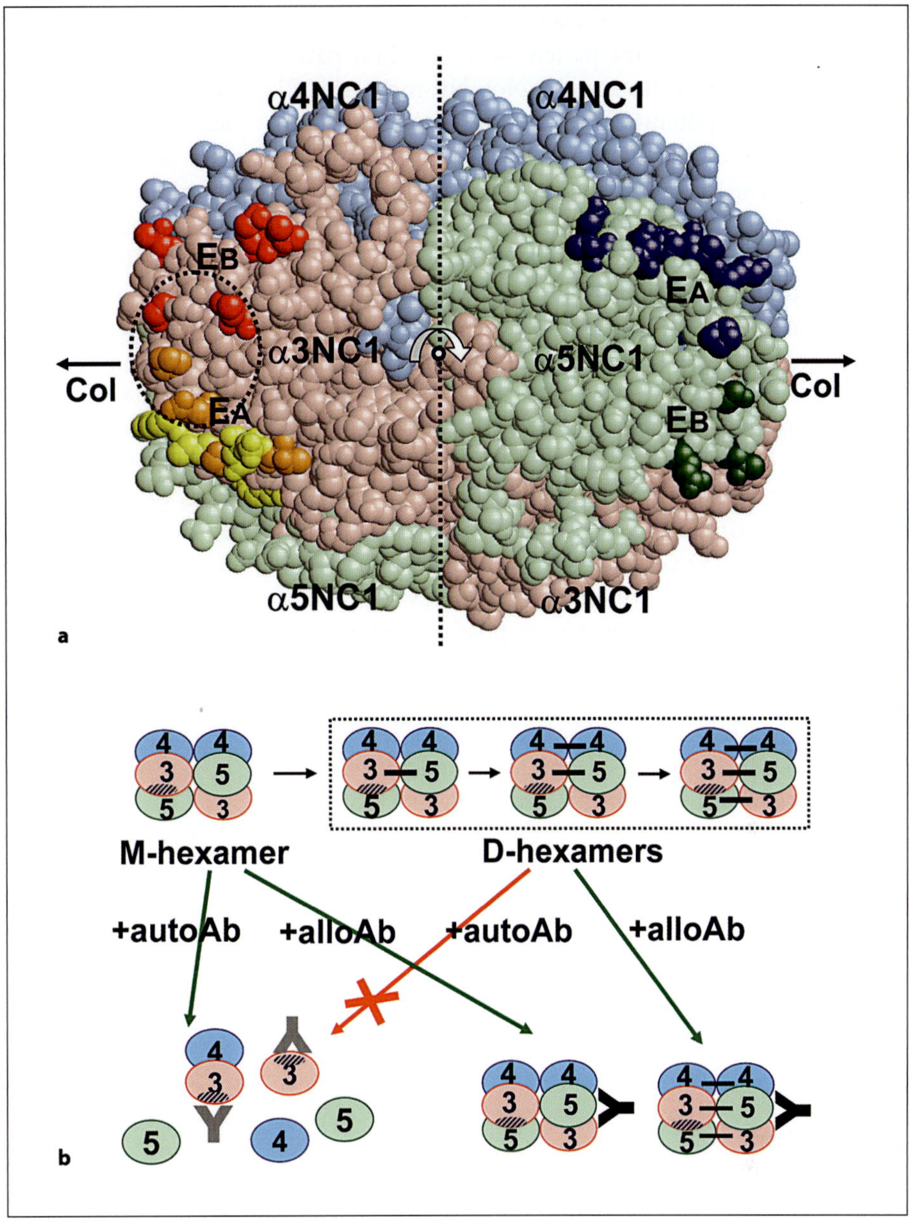

Fig. 1. a A structural representation of the α3α4α5NC1 hexamer depicts the relative location of the putative autoepitope and alloepitope residues within the α3NC1 (light red) and α5NC1 (light green) subunits. The α3NC1-specific residues known to bind autoantibodies are shown in yellow (four residues within the E_A region as well as Gln-57). Other α3NC1 residues in the E_A region are shown in orange, and those in the E_B region in red. Parts of the E_A and E_B regions proximal to one another (dotted circle) encompass the putative epitope of ARAS alloantibodies. The candidate α5NC1 residues potentially important for binding of XLAS alloantibodies are clustered in two regions (dark blue and dark green) that are homologous to the E_A and E_B regions of α3NC1. **b** The differential reactivity of M-α3α4α5 NC1 hexamers (left) and D-α3α4α5 NC1 hexamers (right) with anti-GBM autoantibodies and Alport alloantibodies. The autoepitopes are cryptic in the NC1 hexamer complex, but become accessible once the quaternary structure is breached. Autoantibodies can cause the dissociation of M-hexamers, which lack reinforcing cross-links, but cannot dissociate D-hexamers, which are stabilized by NC1-NC1 cross-links (represented as horizontal lines). In contrast, alloantibodies can bind both M-hexamers and D-hexamers, because their alloepitopes are exposed in the quaternary structure.

bodies recognizing α5NC1 residues 169–229 are present in small amounts in some patient sera, but their pathogenic significance is less certain. Notably, most α5NC1-specific residues potentially important for XLAS alloantibody binding cluster in two short α5NC1 regions, homologous to the E_A and E_B epitope regions of α3NC1 [23]. However, in contrast to the cryptic α3NC1 autoepitopes, the α5NC1 XLAS alloepitopes are accessible on the surface of α3α4α5NC1 hexamers. As a result, alloantibodies can bind to both M- and D-isoforms of α3α4α5 NC1 hexamers, regardless of the presence of NC1 cross-links (fig. 1b). Thus, in Alport post-transplant nephritis, the proximal antigen initiating the alloimmune reaction may be either or both native isoforms of α3α4α5(IV) collagen occurring in the renal allograft, without a specific requirement for unmasking of cryptic epitopes as in autoimmune anti-GBM/GP disease.

Production of alloantibodies is necessary but not sufficient for development of Alport post-transplant anti-GBM nephritis. A significant proportion of transplanted Alport patients have linear IgG deposition in the allograft GBM or circulating anti-GBM alloantibodies to NC1 subunits of α3α4α5(IV) collagen [21], yet do not develop full-blown glomerulonephritis. Although all Alport post-transplant sera exhibit some degree of reactivity to recombinant α3NC1, α4NC1 or α5NC1 domains expressed in *E. coli*, most (>70%) fail to stain kidney sections by immunofluorescence [21]. These alloantibodies may lack nephritogenicity because they target cryptic or misfolded epitopes not present in the GBM. Yet in another ~25% transplanted Alport patients, nonpathogenic alloantibodies have similar properties to pathogenic alloantibodies, including the ability to bind to the GBM, suggesting that additional immune effectors are necessary for the development of glomerulonephritis, and the severity of disease may be modulated by antirejection therapies and/or by additional as yet unknown genetic or environmental factors.

Established post-transplant anti-GBM nephritis is relatively refractory to standard therapy used to treat patients with autoimmune anti-GBM/GP disease, and therefore more proactive interventions would be highly desirable. Increased knowledge of the alloantigen structure, along with the characterization of the pathogenic epitopes recognized by alloreactive B cells and T cells, provide a foundation for the development of novel therapeutic strategies aimed at preemptively inducing immune tolerance in the subset of Alport patients at high risk of developing post-transplant nephritis after a kidney transplant, including those with a history of disease in previous allografts.

Acknowledgements

This work has been supported in part by National Institutes of Health grants P01 DK065123 and R21 AI069874 to D.-B.B.

References

1 Netzer KO, Leinonen A, Boutaud A, Borza DB, Todd P, Gunwar S, Langeveld JP, Hudson BG: The Goodpasture autoantigen: mapping the major conformational epitope(s) of alpha3(IV) collagen to residues 17–31 and 127–141 of the NC1 domain. J Biol Chem 1999;274:11267–11274.
2 Hellmark T, Burkhardt H, Wieslander J: Goodpasture disease: characterization of a single conformational epitope as the target of pathogenic autoantibodies. J Biol Chem 1999;274:25862–25868.
3 Borza DB, Bondar O, Todd P, Sundaramoorthy M, Sado Y, Ninomiya Y, Hudson BG: Quaternary organization of the Goodpasture autoantigen, the α3(IV) collagen chain: sequestration of two cryptic autoepitopes by intra-protomer interactions with the α4 and α5 NC1 domains. J Biol Chem 2002;277:40075–40083.
4 Sundaramoorthy M, Meiyappan M, Todd P, Hudson BG: Crystal structure of NC1 domains: structural basis for type IV collagen assembly in basement membranes. J Biol Chem 2002;277:31142–31153.
5 Than ME, Henrich S, Huber R, Ries A, Mann K, Kuhn K, Timpl R, Bourenkov GP, Bartunik HD, Bode W: The 1.9-A crystal structure of the noncollagenous (NC1) domain of human placenta collagen IV shows stabilization via a novel type of covalent Met-Lys cross-link. Proc Natl Acad Sci USA 2002;99:6607–6612.
6 Chen L, Hellmark T, Wieslander J, Bolton WK: Immunodominant epitopes of alpha3-(IV)NC1 induce autoimmune glomerulonephritis in rats. Kidney Int 2003;64:2108–2120.
7 David M, Borza DB, Leinonen A, Belmont JM, Hudson BG: Hydrophobic amino acid residues are critical for the immunodominant epitope of the Goodpasture autoantigen: a molecular basis for the cryptic nature of the epitope. J Biol Chem 2001;276:6370–6377.
8 Gunnarsson A, Hellmark T, Wieslander J: Molecular properties of the Goodpasture epitope. J Biol Chem 2000;275:30844–30848.
9 Borza DB, Bondar O, Colon S, Todd P, Sado Y, Neilson EG, Hudson BG: Goodpasture autoantibodies unmask cryptic epitopes by selectively dissociating autoantigen complexes lacking structural reinforcement: novel mechanisms for immune privilege and autoimmune pathogenesis. J Biol Chem 2005;280:27147–27154.

10 Vanacore RM, Friedman DB, Ham AJ, Sundaramoorthy M, Hudson BG: Identification of S-hydroxylysyl211-methionine93 as the covalent crosslink of the noncollagenous (NC1) hexamer of the alpha 1alpha 1alpha 2 collagen IV network: a role for the posttranslational modification of lysine211 to hydroxylysine211 in hexamer assembly. J Biol Chem 2005;280:29300–29310.
11 Kalluri R, Cantley LG, Kerjaschki D, Neilson EG: Reactive oxygen species expose cryptic epitopes associated with autoimmune Goodpasture syndrome. J Biol Chem 2000;275: 20027–20032.
12 Wong D, Phelps RG, Turner AN: The Goodpasture antigen is expressed in the human thymus. Kidney Int 2001;60:1777–1783.
13 Salama AD, Chaudhry AN, Ryan JJ, Eren E, Levy JB, Pusey CD, Lightstone L, Lechler RI: In Goodpasture's disease, CD4(+) T cells escape thymic deletion and are reactive with the autoantigen alpha3(IV)NC1. J Am Soc Nephrol 2001;12:1908–1915.
14 Cairns LS, Phelps RG, Bowie L, Hall AM, Saweirs WW, Rees AJ, Barker RN: The fine specificity and cytokine profile of T-helper cells responsive to the alpha3 chain of type IV collagen in Goodpasture's disease. J Am Soc Nephrol 2003;14:2801–2812.
15 Hopfer H, Maron R, Butzmann U, Helmchen U, Weiner HL, Kalluri R: The importance of cell-mediated immunity in the course and severity of autoimmune anti-glomerular basement membrane disease in mice. FASEB J 2003;17:860–868.
16 Kohda T, Okada S, Hayashi A, Kanzaki S, Ninomiya Y, Taki M, Sado Y: High nephritogenicity of monoclonal antibodies belonging to IgG2a and IgG2b subclasses in rat anti-GBM nephritis. Kidney Int 2004;66:177–186.
17 Wu J, Hicks J, Borillo J, Glass WF 2nd, Lou YH: CD4(+) T cells specific to a glomerular basement membrane antigen mediate glomerulonephritis. J Clin Invest 2002;109: 517–524.
18 Wu J, Borillo J, Glass WF, Hicks J, Ou CN, Lou YH: T-cell epitope of alpha3 chain of type IV collagen induces severe glomerulonephritis. Kidney Int 2003;64:1292–1301.
19 Salama AD, Chaudhry AN, Holthaus KA, Mosley K, Kalluri R, Sayegh MH, Lechler RI, Pusey CD, Lightstone L: Regulation by CD25+ lymphocytes of autoantigen-specific T-cell responses in Goodpasture's (anti-GBM) disease. Kidney Int 2003;64:1685–1694.
20 Brainwood D, Kashtan C, Gubler MC, Turner AN: Targets of alloantibodies in Alport anti-glomerular basement membrane disease after renal transplantation. Kidney Int 1998;53:762–766.
21 Kalluri R, Torre A, Shield CF 3rd, Zamborsky ED, Werner MC, Suchin E, Wolf G, Helmchen UM, van den Heuvel LP, Grossman R, Aradhye S, Neilson EG: Identification of alpha3, alpha4, and alpha5 chains of type IV collagen as alloantigens for Alport posttransplant anti-glomerular basement membrane antibodies. Transplantation 2000;69:679–683.
22 Wang XP, Fogo AB, Colon S, Giannico G, Abul-Ezz SR, Miner JH, Borza DB: Distinct epitopes for anti-glomerular basement membrane Alport alloantibodies and Goodpasture autoantibodies within the noncollagenous domain of {alpha}3(IV) collagen: a Janus-faced antigen. J Am Soc Nephrol 2005; 16:3563–3571.
23 Kang JS, Kashtan CE, Turner AN, Heidet L, Hudson BG, Borza DB: The alloantigenic sites of alpha3alpha4alpha5(IV) collagen: pathogenic X-linked Alport alloantibodies target two accessible conformational epitopes in the alpha5NCl domain. J Biol Chem 2007;282:10670–10677.

nephron
Experimental
Nephrology

Nephron Exp Nephrol 2007;106:e44–e50
DOI: 10.1159/000101792

Published online: June 6, 2007

Ontogenetic Development of the Filtration Barrier

Wilhelm Kriz

Anatomy and Cell Biology, Medical Faculty Mannheim, University of Heidelberg, Heidelberg, Germany

Key Words

Glomerular development · Capillary loop stage · Podocyte development · Glomerular developmental defects

Abstract

The development of the filtration barrier is part of a complex sequence of steps proceeding from the early nephron anlage (renal vesicle) via the comma- and S-shaped body to the capillary loop stage and mature glomerulus. The main players are the podocytes (already in the stage of presumptive podocytes), which hold the commander function in this process, and the endothelial and the mesangial cells. A decisive role is also played by the GBM; its change in composition during the developmental process is a precondition for the final maturation of the podocytes, i.e. for the formation of the foot processes and, clearly subsequent, the slit membrane. Failure in the consecutive developmental stages due to genetic mutations or manipulations leads to characteristic hereditary diseases of increasing severity. The last step in this development, the formation of the slit membrane, marks a caesura between diseases with early and late onset; all disorders without a properly developed slit membrane start prenatally or at birth.

The mature filtration barrier consists of a unique porous endothelium, a thick specifically composed basement membrane (glomerular basement membrane; GBM) and the outer layer of interdigitating foot processes, originating from a kind of epithelial pericytes, the podocytes. This particular structure is established in ontogeny along with the development of a renal corpuscle from a nephron anlage [1, 2].

A nephron anlage emerges upon the contact of the ureteral bud with the metanephrogenic blastema. In humans, this occurs in between the third and seventh months of gestation, in rodents it starts prenatally but continues after birth.

At the contact point between both primordia, the nephrogenic blastema transforms into epithelial cells that become separated from the mesenchyme by a basement membrane. This accumulation of cells first takes the shape of a comma (comma-shaped body). After formation of a second indentation, the nephron anlage changes into an S-shaped body which has a proximal and a distal cleft (fig. 1a). Within the distal cleft the precursor cells of podocytes, a disk-like layer of cuboidal epithelial cells, develop opposite to the precursor cells of the macula densa. In the management of this development WT1 and PAX2 appear to play the central role [3]. PAX2 stimulates cell proliferation, i.e. the growth of the nephron

KARGER

Fax +41 61 306 12 34
E-Mail karger@karger.ch
www.karger.com

Accessible online at:
www.karger.com/nee

Dr. Wilhelm Kriz
Anatomy and Cell Biology, INF 307
DE–69120 Heidelberg (Germany)
Tel. +49 6221 54 8680, Fax +49 6221 54 4951
E-Mail wilhelm.kriz@urz.uni-heidelberg.de

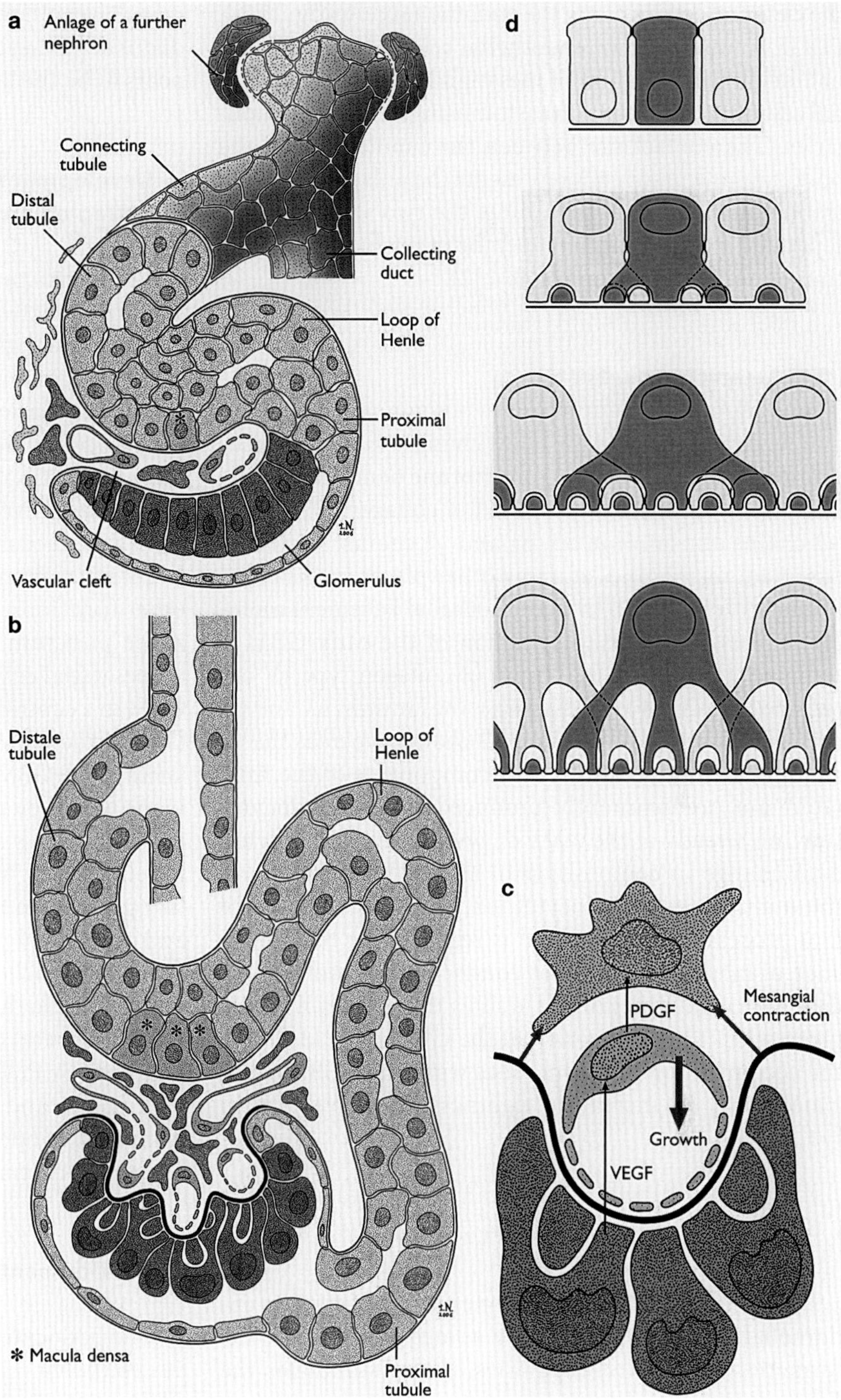

Fig. 1. Stages in the development of a renal glomerulus and filtration barrier. **a** The S-shaped body stage after formation of a distal cleft within the nephron anlage. The cleft separates the presumptive podocytes from the presumptive macula densa cells and is filled by endothelial and mesangial cells. **b** The early capillary loop stage with beginning formation of a glomerular tuft; the central mechanism in this process. **c** The podocyte endothelial cell-mesangial cell axis. **d** The four sketches show consecutive stages in the development of podocytes, which start as simple polygonal epithelial cells connected by an apical junctional belt and terminate as complexly shaped cells with interdigitating foot processes that are connected by a slit membrane (see text for further explanation).

anlage, whereas WT1 gradually suppresses PAX2 expression, i.e. reduces cell proliferation and instead stimulates cell differentiation including the formation and maturation of podocytes. Obviously in response to VEGF A, secreted from the presumptive podocytes downstream to WT1 [4], vascular endothelial cells are attracted to migrate into the distal cleft and to settle in-between the presumptive podocytes and the macula densa cells [5, 6]. Concomitant with the endothelial cells a second group of mesenchymal cells, the mesangial precursor cells, enter the cleft and settle among the endothelial cells (fig. 1a, b). PDGF B secreted from endothelial cells is the major signal for the immigration of mesangial cells [7]. Endothelial cells assemble into an anastomosing network of capillaries. The contact site between the capillaries and the podocyte precursor cells represents the anlage for the filtration barrier; at the beginning, the two structures are separated by two basement membranes which later fuse, thus establishing the early GBM [2].

GBM Development

The composition of the GBM changes during its maturation. The first basement membrane beneath the presumptive podocytes is composed of collagen IV, $\alpha1$ and $\alpha2$ chains and laminin $\alpha1$, $\beta1$ and $\gamma1$ chains, similar to basement membranes at many other places in the body. Along with its fusion to the endothelial basement membrane, this primitive composition of the early GBM is changed; the $\alpha1$ and $\alpha2$ chains of collagen type IV are replaced by $\alpha3$, $\alpha4$ and $\alpha5$ chains, the laminin $\alpha1$ and $\beta1$ chains are replaced by $\alpha5$ and $\beta2$ chains, whereas the $\gamma1$ chain is maintained [8]. These components of the final GBM are predominantly produced by the podocytes starting already at the stage of presumptive podocytes. This change in composition of the GBM accompanies the maturation of podocytes (i.e. for the formation of foot processes; see later) and is required for the adherence of mesangial cells that connect to the laminin $\alpha5$ chain via $\alpha3\beta1$ integrins as well as the Lutheran glycoprotein [9]. The maturation of the GBM coincides with the accumulation of anionic sites within the GBM associated with an increasing tightness to native ferritin [10].

Development of Glomerular Capillaries

VEGF A produced by podocytes not only attracts endothelial cells into the nephron anlage and initiates the formation of the first capillaries, it also stimulates the further growth and sprouting of capillaries. In addition, VEGF is also responsible for the maturation of the glomerular endothelium [5, 6]. The first capillaries encountered at this site have fenestrae with negatively charged diaphragms [10]; mature glomerular capillaries do not have fenestrae bridged by a diaphragm (as is widely believed), instead they have pores lacking a diaphragm. This porous type of capillary is exclusively found in glomerular capillaries. VEGF is certainly not the only mediator of glomerular capillary development [11], but it appears to be the key regulator.

Development of the Mesangium and Extraglomerular Mesangium

Both structures originate from mesenchymal cells that are attracted to move into the distal cleft by PDGF B [7]. The mesangial cells proliferate among the capillaries and form connections to the GBM between the contact sites of capillaries. The prospective extraglomerular mesangial cells assemble at the opposite site making intimate contact with the basement membrane underneath the presumptive macula densa cells (fig. 1). Later, they form a solid cushion of cells between the afferent and efferent arterioles and the basis of the macula densa. Among other functions, the extraglomerular mesangium serves as the 'punctum fixum', as the anchor for the contraction of mesangial cells proper permitting them to pull on the GBM in a centripetal direction [7]. This creates invaginations of the GBM in between the contact points of capillaries to the GBM, thus leads to the formation of capillary loops which bulge into the urinary space. Thereby, the capillary loop stage of glomerular development is established (fig. 1b). It appears that the balance between capillary growth leading to formation of new capillary loops and centripetal pull on the GBM between the capillaries toward the anchor at the macula densa is of fundamental importance in the formation of a glomerular tuft (fig. 1c). The centripetal pull of mesangial cells most probably is regulated by PDGF B. Thus, the coordinated stimulation by VEGF A and PDGF B may represent the key mechanism in development, maintenance and repair of glomerular tuft structure [6, 7, 12].

Development of Podocytes

The podocyte precursor cells form a cuboidal/columnar epithelial cell layer sitting basally on the prospective GBM and being apically connected by a junctional belt (an occluding junction of insufficiently defined variety with features of tight and adherens junctions). Already at an early time point during the formation of the capillary loop stage these presumptive podocytes undergo dramatic changes in cell shape; the cells start to interdigitate with each other leading finally to the foot process pattern and the apical junctional belt moves towards the basal aspect of the cell and transforms into the slit membrane [13] (fig. 1d).

Originally, as is usual in epithelial layers, the junctional belt connects the podocyte precursor cells near the apical surface, separating the apical from the basolateral cell membrane domain. The gradual displacement of the junctional belt towards the basis of the cells [13] creates, in addition to a basolateral cell membrane domain, an apicolateral domain. Concurrently, the cells undergo a complex process of interdigitation starting from the basal aspect of the cells; thus the interdigitation first comprises the basal cell portions (i.e. beneath the junctional belt). This process of basolateral interdigitation is associated with a growth of the basal parts of the cells. At a certain level near the basal aspect of the cells, the junctional belt on its movement towards the base of the cells meets the area of basolateral cell interdigitation. At this border a surprising event occurs: the junctional belt continues its movement down along the interdigitating cell processes. As an important consequence, the apical parts of the interdigitating cell processes change from a basolateral into an apicolateral position. This process continues for some time, displacing major portions of the interdigitating cell processes into an apicolateral position: only the sole plates of the foot processes remain in a basolateral position beneath the junctional belt. Apical membranes, holding true for apicolateral membranes, have free surfaces, and thus are not associated with the corresponding surfaces of an adjacent cell. This allows them to develop individually into processes of different shapes emerging from the cell body and extending in individual ways towards the foot processes (fig. 1d). This mechanism may explain the origin of the primary podocyte cell processes (and any other category of processes above the foot processes) and it explains the lifting of the cell bodies from the GBM, which change into a floating position within the urinary space.

The displacement of the junctional belt from an apical into a basal position, coupled with the change from the demand to interconnect simply shaped polygonal cells to the necessity to circumnavigate thousands of individual processes, requires a dramatic increase in length. Thus, the splitting of the basal cell portions into processes of increasing delicacy and the lengthening of the junctional belt must be narrowly coordinated. Along with this process, the junctional belt undergoes a further dramatic change, this time in composition and architecture. It changes into a unique membrane-like intercellular junction, known as the slit membrane or slit diaphragm. Details of this transition are poorly understood, and it is beyond the scope of this review to consider details of what may happen at the molecular level. If the reader compares the composition of either a tight or an adherens junction with the complex architecture of the slit membrane (discussed in subsequent reviews herein), he/she will easily recognize that the changes are dramatic and that it will take enormous efforts to elucidate the sequence of molecular changes. Parallel to the formation of filtration slits, sialoglycoproteins appear on epithelial surfaces establishing the negatively charged glycocalyx of the podocytes [10].

Defects in Glomerular Development due to Mutations or Genetic Manipulations

Defects in glomerular development are known from a variety of hereditary human diseases as well as from genetic manipulations in mice. The severity of the disease clearly correlates with the developmental stage that fails and the severity of the defect at a certain stage. Even a short look at the symptoms leads to valuable insights.

Formation of the S-Shaped Body Fails

The best known example is the Wilms' tumor emerging from mutations in the *WT1* gene, leading to an impaired balance in the production of the gene products of *PAX2* and *WT1* [14]; a nephron anlage is established but does not undergo proper development. The *Wt1–/–* mouse [15] has no kidneys and develops Wilms' tumor analogies.

Formation of the Glomerular Tuft Fails or Is Impaired

A human example is the Denys Drash syndrome [16, 17], which develops in response to mutations in the *WT1* gene (high risk of a Wilms' tumor). The malformation of glomeruli (small, collapsed) increases in severity in later nephron generations, possibly due to an increasing failure of VEGF A and PDGF B production. The corresponding mutant mice clearly show similar defects: PDGF system knockout mice [7] do not develop any partitions of the tuft, and podocyte-specific *Vegfa–/–* mice ([18]; the total knockout is lethal) die shortly after birth, exhibiting a defective tuft formation with lack of mesangial cells, lack of endothelial pores and defects in slit membrane formation. Other examples for this stage of glomerular malformation are the *Lmx1b–/–* mouse [19, 20], and the *Lama5–/–* mouse [21], which shows that correct GBM development is a precondition for the development of a proper glomerular tuft. Most interesting is a chimeric *Lama5* mutant mouse substituted with the human laminin α1 LG domain [9]; this mouse developed a GBM and intact podocytes and endothelial cells but failed to develop partitions of the tuft due to the inability of mesangial cells to adhere to the abnormal GBM.

Formation of the Foot Processes Fails or Is Impaired

The failure of foot process development is always associated with a failure of slit membrane formation; thus, without foot processes a slit membrane never develops. The following mutant mice all fail to develop foot processes and thus, can be assigned to this category: α3 integrin–/– mouse [22], *Podxl–/–* [23], *Pod1–/–* [24] and probably also the *Fat1–/–* mouse [25].

Formation of a Slit Membrane Fails or Is Severely Impaired

The best example are mutations in the *NPHS1* gene [26, 27], but possibly also some mutations in the *NPHS2* gene [28]. There are several mutant mice that show this damage pattern: the *Nphs1–/–* mouse [29], the nephrin trap/trap mouse [30], the *Nphs2–/–* [31] and the *Neph1–/–* mouse [32]. All these defects in slit membrane composition do not interfere with the process of foot process interdigitation, clearly showing that slit membrane formation is secondary to foot process formation. However, the foot processes under these circumstances, even if initially developed, undergo effacement later probably due to structural and signaling deficiencies in the slit membrane. Thus, the maintenance of the interdigitating foot process pattern is dependent on a proper slit membrane.

A problem in this context is created by the Pierson syndrome, which results from mutations in the *LAMB2* gene [33]. A GBM of normal appearance is developed, but the lack of the laminin β2 chain obviously interferes with the formation of a functioning filtration barrier resulting in a congenital nephrotic syndrome; glomeruli – as seen in published pictures – undergo degeneration along the collapsing pathway. On the other hand, the corresponding mutant *Lamb2–/–* mouse [34, 35], which is much healthier than the *Lama5–/–* mouse [21], exhibits a structurally intact GBM, and the podocytes that contact it do extend foot processes and form a slit membrane, but these are not maintained; this is associated with massive proteinuria shortly after birth. The pathway along which these nephrons finally undergo degeneration is presently unknown.

Foot Processes and the Slit Membrane Develop but Cannot Be Maintained Lifelong (in Models for Only Short Periods)

Under this heading all hereditary glomerular diseases with late onset may be listed, i.e. the different kinds of familial FSGS (*ACTN4* [36], *MYH9* [37], *TRPC6* [38, 39]), Alport, Frasier [40] and Nail-Patella [41] syndromes, and mutations in the *NPHS2* gene with late onset [31]. There are many examples of mutant mice known in which the disease follows this pattern, with delayed onset of disease: *Actn4–/–* [42]; *Synpo–/–* [43]; *Cd2ap–/–* [44]; *Col4a3/Col4a4–/–* [45] and certain manipulations in the *Wt1* gene (+ KTS+/– [46]).

In all these examples the failure of the kidney develops gradually in adult life (in the animal models more rapidly). Thus, not the development, but the maintenance of the glomerular filtration barrier is severely impaired, causing at some time the degeneration of the respective glomerulus. In most of these mutations with late onset the degeneration of the glomerulus follows the pathway to 'classic' FSGS which is based on a progressive depletion of podocytes. It has been suggested that under such conditions an increased vulnerability of podocytes to external, even physiological challenges (e.g. variations in glomerular pressure) acting as a 'second hit' [47] may underlie a comparably high frequency of podocyte loss. Since lost podocytes cannot be replaced by cell replication [see Marshall and Shankland, this issue], the general pathway to 'classic' FSGS becomes effective: too few podocytes finally result in FSGS and loss of individual nephrons; too few nephrons finally lead to renal failure [48].

This secondary mechanism of nephron loss is clearly different from nephron degeneration starting at developmental stages in which the disease manifests already with an erroneous development of the filtration barrier before or at birth. In these cases the degeneration of the glomerulus follows the collapsing pathway (or a variant of it) as has recently been shown for *NPHS1* by Kuusniemi et al. [27] fully agreeing with own unpublished observations.

References

1 Saxen L, Sariola H, Lehtonen E: Sequential cell and tissue interactions governing organogenesis of the kidney. Anat Embryol 1986; 175:1–6.

2 Abrahamson D: Glomerulogenesis in the developing kidney. Semin Nephrol 1991;11: 375–389.

3 Discenza M, He S, Lee T, Chu L, Bolon B, Goodyer P, Eccles M, Pelletier J: WT1 is a modifier of the Pax2 mutant phenotype: cooperation and interaction between WT1 and Pax2. Oncogene 2003;22:8145–8155.

4 Hanson J, Gorman J, Reese J, Fraizer G: Regulation of vascular endothelial growth factor, VEGF, gene promoter by the tumor suppressor, WT1. Front Biosci 2007;1: 2279–2290.
5 Eremina V, Haigh J, Nagy A, Kikkawa Y, Miner J, Quaggin S: The role of VEGF-A in glomerular angiogenesis. J Am Soc Nephrol 2002;13:100A.
6 Eremina V, Quaggin S: The role of VEGF-A in glomerular development and function. Curr Opin Nephrol Hypertens 2004;13:9–15.
7 Betzholtz C, Lindblom P, Bjarnegard M, Enge M, Gerhardt H, Lindahl P: Role of platelet-derived growth factor in mesangium development and vasculopathies: lessons from platelet-derived growth factor and platelet-derived growth factor receptor mutations in mice. Curr Opin Nephrol Hypertens 2005;13:45–52.
8 Miner JH: Building the glomerulus: a matricentric view. J Am Soc Nephrol 2005;16:857–861.
9 Kikkawa Y, Virtanen I, Miner JH: Mesangial cells organize the glomerular capillaries by adhering to the G domain of laminin alpha5 in the glomerular basement membrane. J Cell Biol 2003;161:187–196.
10 Reeves W, Kanwar Y, Farquhar M: Assembly of the glomerular filtration surface: differentiation of anionic sites in glomerular capillaries of newborn rat kidney. J Cell Biol 1980; 85:735–753.
11 Abrahamson D, Robert B: Derivation and differentiation of glomerular endothelial cells. Nephrol Dial Transplant 2003;18:vi2–vi7.
12 Kriz W, Haehnel B, Hosser H, Ostendorf T, Kränzlin B, Gretz N, Shimizu F, Floege J: Pathways to recovery and loss of nephrons in anti-Thy-1 nephritis. J Am Soc Nephrol 2003;14:1904–1926.
13 Reeves W, Caulfield J, Farquhar M: Differentiation of epithelial foot processes and filtration slits: sequential appearance of occluding junctions, epithelial polyanion, and slit membranes in developing glomeruli. Lab Invest 1978;39:90–100.
14 Hastie N: The genetics of Wilms' tumor: a case of disrupted development. Annu Rev Genet 1994;28:523–558.
15 Kreidberg J, Sariola H, Loring J, Maeda M, Pelletier J, Housman D, Jaenisch R: WT-1 is required for early kidney development. Cell 1993;74:679–691.
16 Habib R, Loirat C, Gubler M, Niaudet P, Bensman A, Levy M, Broyer M: The nephropathy associated with male pseudohermaphroditism and Wilms' tumor. Drash syndrome: a distinctive glomerular lesion – report of 10 cases. Clin Nephrol 1985;24: 269–278.
17 Yang Y, Jeanpierre C, Dressler G, Lacoste M, Nidaudet P, Gubler M: WT1 and PAX-2 podocyte expression in Denys-Drash syndrome and isolated diffuse mesangial sclerosis. Am J Pathol 1999;154:181–192.
18 Eremina V, Sood M, Haigh J, Nagy A, Lajoie G, Ferrara N, Gerber HP, Kikkawa Y, Miner JH, Quaggin SE: Glomerular-specific alterations of VEGF-A expression lead to distinct congenital and acquired renal diseases. J Clin Invest 2003;111:707–716.
19 Miner JH, Morello R, Andrews KL, Li C, Antignac C, Shaw AS, Lee B: Transcriptional induction of slit diaphragm genes by Lmx1b is required in podocyte differentiation. J Clin Invest 2002;109:1065–1072.
20 Rohr C, Prestel J, Heidet L, Hosser H, Kriz W, Johnson R, Antignac C, Witzgall R: The LIM-homeodomain transcription factor Lmx1b plays a crucial role in podocytes. J Clin Invest 2002;109:1073–1082.
21 Miner JH, Li C: Defective glomerulogenesis in the absence of laminin alpha5 demonstrates a developmental role for the kidney glomerular basement membrane. Dev Biol 2000;217:278–289.
22 Kreidberg J, Donovan M, Goldstein S, Rennke H, Shepherd K, Jones R, Jaenisch R: Alpha 3 beta 1 integrin has a crucial role in kidney and lung organogenesis. Development 1996; 122:3537–3547.
23 Doyonnas R, Kershaw D, Duhme C, Merkens H, Chelliah S, Graf T, McNagny K: Anuria, omphalocele, and perinatal lethality in mice lacking the CD34-related protein podocalyxin. J Exp Med 2001;194:13–27.
24 Quaggin S, Schwartz L, Cui S, Igarashi P, Deimling J, Post M, Rossant J: The basic-helix-loop-helix protein Pod1 is critically important for kidney and lung organogenesis. Development 1999;126:5771–5783.
25 Ciani L, Patel A, Allen N, ffrench-Constant C: Mice lacking the giant protocadherin mFAT1 exhibit renal slit junction abnormalities and a partially penetrant cyclopia and anophthalmia phenotype. Mol Cell Biol 2003;23:3575–3582.
26 Kestila M, Lenkkeri U, Männikkö M, Lamerdin J, McCready P, Putaala H, Ruotsalainen V, Morita T, Nissinen M, Herva R, Kashtan C, Peltonen L, Holmberg C, Olsen A, Tryggvason K: Positionally cloned gene for a novel glomerular protein – nephrin – is mutated in congenital nephrotic syndrome. Mol Cell 1998;1:575–582.
27 Kuusniemi A-M, Merenimes J, Lahdenkari A-T, Holmberg C, Salmela K, Karikoski R, Rapola J, Jalanko H: Glomerular sclerosis in kidneys with congenital nephrotic syndrome (NPHS1). Kidney Int 2006;70:1423–1431.
28 Boute N, Gribouval O, Roselli S, Benessy F, Lee H, Fuchshuber A, Dahan K, Gubler M, Niaudet P, Antignac C: NPHS2, encoding the glomerular protein podocin, is mutated in autosomal recessive steroid-resistant nephrotic syndrome. Nat Genet 2000;24:349–354.
29 Putaala H, Soininen R, Kilpelainen P, Wartiovaara J, Tryggvason K: The murine nephrin gene is specifically expressed in kidney, brain and pancreas: inactivation of the gene leads to massive proteinuria and neonatal death. Hum Mol Genet 2001;10:1–8.
30 Rantanen M, Palmen T, Patari A, Ahola H, Lehtonen S, Astrom E, Floss T, Vauti F, Wurst W, Ruiz P, Kerjaschki D, Holthofer H: Nephrin TRAP mice lack slit diaphragms and show fibrotic glomeruli and cystic tubular lesions. J Am Soc Nephrol 2002;13:1586–1594.
31 Roselli S, Heidet L, Sich M, Henger A, Kretzler M, Gubler M, Antignac C: Early glomerular filtration defect and severe renal disease in podocin-deficient mice. Mol Cell Biol 2005;24:550–560.
32 Donoviel D, Freed D, Vogel H, Potter D, Hawkins E, Barrish J, Mathur B, Turner C, Geske R, Montgomery C, Starbuck M, Brandt M, Gupta A, Ramirez-Soliz R, Zambrowicz B: Proteinuria and perinatal lethality in mice lacking NEPH1, a novel protein with homology to nephrin. Mol Cell Biol 2001;21:4829–4836.
33 Zenker M, Aigner T, Wendler O, Tralau T, Müntefering H, Fenski R, Pitz S, Schumacher V, Royer-Pokora B, Wühl E, Cochat P, Bouvier R, Kraus C, Mark K, Madlon H, Dötsch J, Rascher W, Maruniak-Chudek I, Lennert T, Neumann L, Reis A: Human laminin β2 deficiency causes congenital nephrosis with mesangial sclerosis and distinct eye abnormalities. Hum Mol Genet 2004;13: 2625–2632.
34 Noakes PG, Miner JH, Gautam M, Cunningham JM, Sanes JR, Merlie JP: The renal glomerulus of mice lacking s-laminin/laminin beta 2: nephrosis despite molecular compensation by laminin beta 1. Nat Genet 1995;10: 400–406.
35 Jarad G, Cunningham J, Shaw AS, Miner JH: Proteinuria precedes podocyte abnormalities in *Lamb2-/-* mice, implicating the glomerular basement membrane as an albumin barrier. J Clin Invest 2006;116:2272–2279.
36 Kaplan C, Kim S, North K, Rennke H, Correia L, Tong H, Mathis B, Rodríguez-Pérez J, Allen P, Beggs A, Pollak M: Mutations in ACTN4, encoding alpha-actinin-4, cause familial focal segmental glomerulosclerosis. Nat Genet 2000;24:251–256.
37 Arrondel C, Vodovar N, Knebelmann B, Grunfeld J, Gubler M, Antignac C, Heidet L: Expression of the nonmuscle myosin heavy chain IIA in the human kidney and screening for MYH9 mutations in Epstein and Fechtner syndromes. J Am Soc Nephrol 2002;13:65–74.
38 Reiser J, Polu K, Moller C, Kenlan P, Altintas M, Wei C, Faul C, Herbert S, Villegas I, Avila-Casado C, McGee M, Sugimoto H, Brown D, Kalluri R, Mundel P, Smith P, Clapham D, Pollak M: TRPC6 is a glomerular slit diaphragm-associated channel required for normal renal function. Nat Genet 2005;37: 739–744.

39 Winn M, Conlon P, Lynn K, Farrington M, Creazzo T, Hawkins A, Daskalakis N, Kwan S, Ebersviller S, Burchette J, Pericak-Vance M, Howell D, Vance J, Rosenberg P: A mutation in the TRPC6 cation channel causes familial focal segmental glomerulosclerosis. Science 2005;308:1801–1804.
40 Barbaux S, Niaudet P, Gubler M, Grunfeld J, Jaubert F, Kuttenn F, Fekete C, Souleyreau-Therville N, Thibaud E, Fellous M, McElreavey K: Donor splice-site mutations in WT-1 are responsible for Frasier syndrome. Nat Genet 1997;17:467–470.
41 Bennett W, Musgrave J, Campbell R, Elliot D, Cox R, Brooks R, Lovrien E, Beals R, Porter G: The nephropathy of the nail-patella syndrome: clinicopathologic analysis of 11 kindred. Am J Med 1973;54:304–319.
42 Kos C, Le T, Sinha S, Henderson J, Kim S, Sugimoto H, Kalluri R, Gerszten R, Pollak M: Mice deficient in alpha-actinin-4 have severe glomerular disease. J Clin Invest 2003; 111:1683–1690.
43 Asanuma K, Kim K, Oh J, Giardino L, Chabanis S, Faul C, Reiser J, Mundel P: Synaptopodin regulates the actin-bundling activity of alpha-actinin in an isoform-specific manner. J Clin Invest 2005;1:1.
44 Shih N-Y, Li J, Karpitskii V, Nguyen A, Dustin ML, Kanagawa O, Miner JH, Shaw AS: Congenital nephrotic syndrome in mice lacking CD2-associated protein. Science 1999;286:312–315.
45 Kashtan C: Animal models of Alport syndrome. Nephrol Dial Transplant 2002;17: 1359–1361.
46 Hammes A, Guo J, Lutsch G, Leheste J, Landrock D, Ziegler U, Gubler M, Schedl A: Two splice variants of the Wilms' tumor 1 gene have distinct functions during sex determination and nephron formation. Cell 2001;106:319–329.
47 Yao J, Le T, Kos C, Henderson J, Allen P, Denker B, Pollak M: Alpha-actinin-4-mediated FSGS: an inherited kidney disease caused by an aggregated and rapidly degraded cytoskeletal protein. PLoS Biology 2004;2:0787–0794.
48 Kriz W: TRPC6 – a new podocyte gene involved in focal segmental glomerulosclerosis. Trends Mol Med 2005;11:527–530.

nephron
Experimental
Nephrology

Nephron Exp Nephrol 2007;106:e51–e59
DOI: 10.1159/000101793

Published online: June 6, 2007

Cell Cycle Regulatory Proteins in Podocyte Health and Disease

Caroline B. Marshall Stuart J. Shankland

Division of Nephrology, University of Washington, Seattle, Wash., USA

Key Words
Cell cycle · Cyclins · CDK inhibitors · Glomerulus · Podocyte · Hypertrophy · Apoptosis · Proliferation · DNA damage

Abstract
The glomerular visceral epithelial cell, or podocyte, is a highly specialized and terminally differentiated cell that is fundamental to the integrity of the glomerular filtration barrier and functions to prevent urinary protein leakage and to oppose intracapillary hydrostatic pressure. Common to many human kidney diseases and experimental animal models is a strong association between podocyte injury and the development of progressive kidney disease. Studies have shown that a decline in podocyte number strongly correlates with, and likely underlies, proteinuria and the progression to glomerulosclerosis. Maintenance of podocyte differentiation, essential to its normal structure and function, is challenged in the setting of glomerular injury, with very divergent outcomes dependent upon the inciting injury. In response to injury, podocytes may undergo several cell fates, including proliferation, de-differentiation, hypertrophy, apoptosis, or necrosis. Common to these potential outcomes of renal injury is their ultimate regulation at the level of the cell cycle. Positive regulators (cyclins and cyclin-dependent kinases) and negative regulators (cyclin-dependent kinase inhibitors) coordinate the cell cycle. There is now a large body of literature confirming the importance of cell cycle regulatory proteins in the cellular response to injury. Emerging lessons from mouse knockout experiments highlight that the cell cycle machinery operates differently in distinct cell types. Recent studies focusing on the roles of cell cycle regulatory proteins specifically in podocytes have provided important clues on how these proteins operate to constrain cell proliferation and preserve differentiation in health, and how they modulate the dysregulated phenotype in diseased states. In disease, both a failure to regenerate lost podocytes and an inappropriate proliferative response can have profound consequences for glomerular structure and function. Here, we will review the latest advances in understanding the roles of cell cycle regulatory proteins in diseases of the podocyte.

We apologize to those whose work could not be cited due to space limitations. An unabridged list of references is available from the authors upon request.

KARGER
Fax +41 61 306 12 34
E-Mail karger@karger.ch
www.karger.com

Accessible online at:
www.karger.com/nee

Stuart J. Shankland, MD
Division of Nephrology, University of Washington
1959 NE Pacific Street, Box 356521
Seattle, WA 98195 (USA)
Tel. +1 206 543 3792, Fax +1 206 685 8661, E-Mail stuartjs@u.washington.edu

Introduction

The glomerular visceral epithelial cell, also called the podocyte, has a highly complex cytoarchitecture that determines its elaborate function as a critical component of the glomerular filtration barrier. Podocytes prevent urinary protein leakage and provide structural support to the capillary wall, thereby counteracting the hydrostatic pressure within the glomerular capillaries. There is very little cell turnover in the mature adult kidney under normal physiological conditions, and the mechanisms governing podocyte differentiation, and therefore proliferation, are essential to the normal function of this cell. As will be discussed elsewhere in this edition, the podocyte is the target of injury in a variety of kidney diseases, including diabetic nephropathy, focal segmental glomerulosclerosis (FSGS), minimal change disease, and membranous nephropathy. As such, the podocyte's complex structure and function may be disturbed by an array of offenders, including toxins, immune-mediated stresses, metabolic derangements, infections, hemodynamic changes, growth factors, and reactive oxygen species.

Regardless of the initiating factor(s), the podocyte's mature and quiescent phenotype is maintained in most diseases, even at the expense of developing foot process effacement and progressive scarring (glomerulosclerosis). Nevertheless, maintenance of podocyte differentiation is challenged in the setting of glomerular injury, and the inciting event determines how the podocyte responds to injury. In fact, in response to injury, podocytes may undergo one or more cell fate(s), including de-differentiation and proliferation, persistent cell cycle arrest, hypertrophy, apoptosis, or necrosis. Although these events may be governed at numerous levels, common to the potential consequences of podocyte injury is their ultimate regulation at the level of the cell cycle. This review will be restricted to a discussion on cell cycle regulatory proteins and their role in podocyte biology, both in health and in disease.

Overview: The Cell Cycle

Although first described and identified for their roles as coordinators of cell cycle progression and proliferation, recent research has characterized a more extensive involvement of cell cycle regulatory proteins in other fundamental biologic processes, including development, differentiation, hypertrophy, and apoptosis. A comprehensive discussion of cell cycle biology is beyond the scope of this review, and the reader is directed to the suggested reference [1].

In brief, playing important roles in cell cycle control are multi-protein complexes consisting of a catalytic subunit (the kinase) and a regulatory subunit (the cyclin). Sequential activation of cyclin/cyclin-dependent kinase (Cdk) complexes regulates passage through the cell cycle, including progression from the quiescent phase (G_0) through DNA synthesis (S phase) and mitosis (M phase). Active cyclin/Cdk complexes phosphorylate and inactivate retinoblastoma protein, leading to induction of E2F-regulated gene expression and cell proliferation. Cdk inhibitors bind and inhibit the activity of cyclin/Cdk complexes and negatively regulate cell cycle progression. Two families of Cdk inhibitors have been identified, on the basis of sequence homology and their target cyclin-Cdk complexes. The INK4 family of Cdk inhibitors specifically targets the G_1-phase cyclin/Cdk complexes, and includes $p15^{INK4b}$, $p16^{INK4a}$, $p18^{INK4c}$, and $p19^{INK4d}$. The Cip/Kip family of Cdk inhibitors inhibits both G_1- and S-phase cyclin/Cdk complexes and includes $p21^{Cip1}$, $p27^{Kip1}$, and $p57^{Kip2}$.

Podocyte Proliferation and De-Differentiation

Podocyte Development: Onset of Differentiation and Quiescence

During normal glomerulogenesis, immature and undifferentiated podocyte precursor cells proliferate. In the S-shaped body stage of glomerulogenesis, presumptive podocytes are engaged in the cell cycle and thus express proliferative tissue markers, including proliferating cell nuclear antigen (PCNA) and Ki-67, together with cyclin A (required for DNA synthesis) and cyclin B1 and Cdc2 (required for mitosis). In both mouse and human glomerulogenesis, immunostaining for the Cdk inhibitors $p27^{Kip1}$ and $p57^{Kip2}$ is absent at this early stage. With transition to the capillary loop stage, a fundamental phenotypic switch occurs with cessation of mitotic activity, reorganization of the cytoskeleton, expression of podocyte maturity markers (including synaptopodin and glomerular epithelial protein-1), and development of foot-process interdigitation. In parallel to these changes, there is a decrease in levels of cell cycle promoters (cyclins and Cdks), and a reciprocal upregulation of Cdk inhibitors $p27^{Kip1}$ and $p57^{Kip2}$. Similarly, levels for $p21^{Cip1}$, $p27^{Kip1}$, and $p57^{Kip2}$ increase in differentiated podocytes grown in culture. These cell cycle events ultimately lead to cell cycle arrest and the cessation of proliferation, and coincide with the establishment of the mature podocyte's terminally differentiated and quiescent phenotype – necessary requirements for normal structure and function.

Podocyte Maturation: Terminal Differentiation and Inability to Proliferate

An important question is what happens to the podocyte at the level of the cell cycle following injury and in states of disease. Studies have shown that following injury, podocytes do occasionally re-enter the cell cycle. Despite this, podocytes typically do not proliferate in most diseased states, including diabetic nephropathy, FSGS, minimal change disease, and membranous nephropathy. The question that follows is whether podocytes lack the machinery required to progress through the cell cycle (cyclins and Cdks) or if there is an excess of cell cycle inhibitors (Cdk inhibitors) blocking cell cycle progression. Although podocytes do express cyclins following injury, albeit at low levels, any Cdk activity is limited by the robust expression and upregulation of the Cdk inhibitors $p21^{Cip1}$, $p27^{Kip1}$, and $p57^{Kip2}$. This upregulation precedes cell cycle arrest at G_1/S phase (limiting DNA synthesis) and G_2/M phase (limiting mitosis). Taken together, in the setting of disease-induced podocyte loss by apoptosis or detachment, the apparent lack of proliferation in response to injury, due to changes within the cell cycle, further contributes to reduced podocyte number.

Podocyte De-Differentiation and Proliferation: Recapitulation of the Fetal Phenotype

Proliferation and the state of differentiation are closely linked in podocytes. Although podocytes don't proliferate in the major causes of nephrotic syndrome (diabetic nephropathy, FSGS, membranous nephropathy, and minimal change disease), idiopathic collapsing glomerulopathy, HIV-associated nephropathy, and crescentic glomerulonephritis are characterized by the dysregulation of the differentiated phenotype. Moreover, this dysregulation is accompanied by, and may lead to, podocyte proliferation. This raises the fundamental question of why podocytes proliferate in certain diseases but not in others. In collapsing glomerulopathy, studies by Shankland et al. [2] and Barisoni et al. [3] showed that podocyte number increased, with associated de-differentiation (loss of maturity markers, including synaptopodin) and phenotypic dysregulation (loss of Wilms' tumor protein). In proliferating podocytes overlying areas of collapse, immunostaining for $p57^{Kip2}$ and $p27^{Kip1}$ was segmentally absent in a distribution nearly identical to that of Ki-67 expression, accompanied by increased cyclin A.

Although the ability of HIV to induce dysregulated podocyte proliferation has been recognized for some time, only recently have studies focused on the direct interaction between specific viral antigens, including Tat and Nef, and cell cycle regulatory proteins. HIV-infected podocytes have increased levels of cyclin D1 and phosphorylated (inactive) retinoblastoma protein, while podocytes infected with HIV virus deleted of Nef do not show induction of cyclin D1 [4]. These data suggest that the HIV virus itself directly interacts with cell cycle regulatory proteins leading to the dysregulated podocyte proliferation seen in HIV-associated nephropathy.

More recently, using somatic Cre recombination constitutively to express β-galactosidase in podocytes in vivo, Moeller et al. [5] have demonstrated that podocytes also actively proliferate in acute cellular crescentic glomerulonephritis. The cells of podocyte origin did not express specific markers of mature podocytes, including nephrin, CD2AP, and podocin, suggesting that these cells had reverted to a less differentiated state. These forms of nonreparative proliferation, including collapsing FSGS, HIV-associated nephropathy, and crescentic glomerulonephritis, are characterized by the podocyte's escape from cell cycle blockade, leading to a disruption of glomerular architecture followed by rapidly declining renal function. Proof of principle for this was evident when we showed recently that inhibiting the activity of Cdk2 with roscovitine significantly reduced podocyte proliferation in experimental disease, which was accompanied by improved renal function and less glomerular scarring.

Taken together, these studies highlight that when the differentiated phenotype is maintained, the podocyte typically remains arrested in the cell cycle and does not proliferate. However, under certain circumstances, podocytes can de-differentiate, allowing re-entry into the cell cycle and subsequent proliferation. Regardless, these events are tightly regulated by specific cell cycle regulatory proteins, which ultimately may determine the podocyte's proliferative phenotype.

Podocyte DNA Damage and Lack of Proliferation

At the cell cycle level, reduced proliferation is a consequence of either an abnormality in DNA synthesis, secondary to G_1/S checkpoint arrest, or reduced mitosis, secondary to G_2/M checkpoint arrest. Below, we review the evidence for abnormalities in DNA synthesis and mitosis in the podocyte's response to injury.

Abnormalities in DNA Synthesis

The passive Heymann nephritis (PHN) model, induced by the administration of an antibody reactive against the rat podocyte Fx1A antigen, has many similarities to human membranous nephropathy. In both diseases, complement activation, with the assembly of the terminal complement components to form the membrane attack complex C5b-9, is the principal mediator of podocyte injury. Nucleated cells, such as podocytes, are relatively resistant to C5b-9 attack and may actually become activated by it, leading to release of inflammatory mediators, calcium, oxidants, growth factors, matrix components, and proteases. C5b-9 is also central to mesangial cell injury in mesangial proliferative glomerulonephritis. In mesangial cells, sublytic C5b-9 attack leads to DNA synthesis and a marked proliferative response. In contrast, although podocytes also re-engage the cell cycle following sublytic C5b-9 injury, as demonstrated by an increase in cyclins and Cdks required for proliferation (cyclin A and Cdk2 protein levels increase), they undergo limited DNA synthesis, and podocyte number typically remains constant, and even decreases with time. These findings are consistent with a defect in DNA synthesis (S phase of cell cycle). Indeed, the levels of $p21^{Cip1}$ and $p27^{Kip1}$ increase specifically in podocytes in experimental membranous nephropathy, supporting the notion that increased Cdk inhibitors prevent podocyte cell cycle progression, and therefore, DNA synthesis [6].

Further support of a critical role for $p21^{Cip1}$ and $p27^{Kip1}$ in limiting the proliferative response of podocytes in many disease processes comes from studies in $p21^{Cip1}$–/– and $p27^{Kip1}$–/– mice. With the administration of an anti-glomerular antibody, podocyte de-differentiation and proliferation increased in $p21^{Cip1}$–/– and $p27^{Kip1}$–/– mice compared with wild-type mice [7]. Podocyte proliferation was accompanied by extracellular matrix accumulation with associated decline in renal function. In addition, inhibiting podocyte proliferation with the potent Cdk inhibitor roscovitine led to reduced glomerulosclerosis and improved renal function in experimental crescentic glomerulonephritis and HIV-associated nephropathy. The roscovitine studies offer further substantiation that strong and specific upregulation of Cdk inhibitors maintains tight control of cell cycle quiescence in podocytes.

Abnormalities in Mitosis: Role of DNA Damage

We stated earlier that abnormalities in DNA synthesis constitute one mechanism for the podocyte's inability to proliferate in disease. An important question that follows is whether abnormalities in podocyte mitosis can account for additional or alternate mechanisms for post-injury limitation of proliferation. Previously, we have shown that in experimental membranous nephropathy, podocytes do increase their levels of cyclin B-Cdc2, the cell cycle proteins required for mitosis. However, mitosis is infrequent. This suggests an abnormality in the G_2/M checkpoint, which would then limit the cell's entry into mitosis. Indeed, sublytic C5b-9 injury causes DNA damage in podocytes in vitro and in vivo, accompanied by a delay or inhibition in entering mitosis, consistent with G_2/M phase arrest [8]. In contrast, upon exposure to sublytic C5b-9, DNA damage is not seen in mesangial cells. Thus, podocytes are uniquely susceptible to genotoxic stress by sublytic C5b-9. The consequences of DNA damage include activation of specific cell cycle checkpoints, molecular pathways that monitor passage through the cell cycle and generate pauses in cell cycle progression when DNA damage is present. Accordingly, DNA damage leads to cell cycle arrest, permitting the cell sufficient time for DNA repair or other cell fate decisions, including apoptosis or entry into a senescence-like state, thereby limiting proliferation.

In our quest to determine the mechanisms of this effect, we found that sublytic C5b-9-induced podocyte DNA damage was associated with a marked increase in cell cycle checkpoint proteins checkpoint kinase-1 (Chk-1), checkpoint kinase-2 (Chk-2), tumor suppressor p53, and $p21^{Cip1}$. These inhibitory proteins converge on and block G_2/M phase progression, thus limiting mitosis and subsequent proliferation [8]. These observations support the hypothesis that $p21^{Cip1}$, a downstream effector of p53 in the checkpoint pathway, plays an essential role in limiting podocyte proliferation. Therefore, we utilized $p21^{Cip1}$–/– mouse podocytes in culture to determine more precisely the role of $p21^{Cip1}$ in maintaining growth arrest in the setting of genotoxic stress. Following exposure of $p21^{Cip1}$–/– podocytes to sublytic C5b-9, there were increased levels of DNA damage compared to that observed in $p21^{Cip1}$+/+ podocytes. This greater degree of DNA damage in $p21^{Cip1}$–/– podocytes was not accompanied by an increase of DNA damage checkpoint proteins (including Chk-1, Chk-2, p53, and GADD45α [growth arrest and DNA damage inducible-45α]) as was seen in the wild-type podocytes [unpubl. data]. Taken together, $p21^{Cip1}$ likely plays a critical role in the podocyte's cell cycle arrest in response to DNA damage and in the regulation of DNA damage response effectors.

DNA damage may also account for the inability of podocytes to proliferate in response to injury in other podocyte diseases. For example, in the puromycin aminonucleoside (PA) nephropathy model of minimal-change disease progressing to FSGS, PA induced oxidant-mediated DNA damage in podocytes. Moreover, this was accompanied by an increase in cell cycle checkpoint proteins p53 and $p21^{Cip1}$. PA also induced DNA damage in cultured mouse podocytes, accompanied by an upregulation of $p21^{Cip1}$, GADD45α, p53, Chk-1, Chk-2, and the DNA repair enzymes DNA polymerase-β and AP-endonuclease [9]. Hence, in both immune- and non-immune-mediated podocyte injury, in vitro and in vivo, DNA damage leading to activation of inhibitory checkpoint pathways, to arrest or delay cell cycle progression, may be a mechanism underlying the podocyte's relative inability to proliferate following injury. In combination with injury-induced podocyte loss from other mechanisms, including apoptosis and detachment, the relative lack of podocyte proliferation, due to DNA damage and the consequent G_2/M phase arrest and reduced mitosis, may contribute to decreased podocyte number and areas of denuded basement membrane. Ultimately, this leads to development of glomerulosclerosis and progressive renal impairment.

Podocyte Hypertrophy

As stated earlier, podocytes in adult kidneys have exited the cell cycle and are quiescent, resting in the G_0 phase. A frequent response to podocyte injury is an increase in cell size due to hypertrophy. This is not to be confused with an increase in cell number, which is a consequence of proliferation. Studies suggest that hypertrophy is an active process in which cells re-enter the cell cycle from G_0. Entry into G_1 is associated with an increase in RNA and protein synthesis, in anticipation of DNA replication in the S phase. However, cells undergoing hypertrophy do not progress into S phase, but instead arrest in G_1. Therefore, the ratio of protein to DNA is increased, leading to increased size (hypertrophy). Recent studies have shown that podocytes may undergo hypertrophy following injury, which likely represents the podocyte's attempt to cover denuded areas of glomerular basement membrane following the loss of neighboring cells due to detachment or apoptosis. However, with fewer podocytes to maintain glomerular capillary loop integrity and oppose intracapillary hydrostatic pressure, the initial adaptive response of podocyte hypertrophy becomes maladaptive as the remaining podocytes perform more work and are subjected to augmented mechanical stretch due to increased intraglomerular capillary pressure, also called glomerular hypertension.

Applying mechanical stretch, the end result of glomerular hypertension, to cultured mesangial cells activates a variety of pro-proliferative signaling pathways, resulting in proliferation. In contrast, we have shown that mechanical stretch decreases podocyte proliferation. We explored the mechanisms of this effect and showed that cultured podocytes exposed to mechanical stretch exhibited decreased levels of cyclins D1, A, and B1, and Cdc2, and increased levels of Cdk inhibitors $p21^{Cip1}$ and $p27^{Kip1}$. Interestingly, exposing cultured mouse podocytes to cyclic mechanical stretch not only inhibits proliferation, but also induces a hypertrophic phenotype [10]. The underlying mechanism may be due to $p21^{Cip1}$. In contrast to the growth arrest and hypertrophy induced in wild-type cells exposed to stretch, $p21^{Cip1}$–/– podocytes continued to proliferate and did not hypertrophy. These studies are consistent with the notion that mechanical stretch inhibits podocyte proliferation and prompts the cell to adopt a hypertrophic phenotype, both events regulated by the Cdk-inhibitor $p21^{Cip1}$, which may explain, in part, why podocytes typically do not proliferate in states of increased intraglomerular pressure.

Diabetic nephropathy is characterized by glomerular hypertrophy. In Zucker diabetic fatty rats (ZDF-*fa/fa*), a model of type 2 diabetes, Hoshi et al. demonstrated increased glomerular volume and sclerosis that was associated with increased expression of $p27^{Kip1}$ and $p21^{Cip1}$, predominantly localized to podocytes, but unchanged levels of $p57^{Kip2}$ [11]. An in vitro correlate confirmed that podocytes exposed to hyperglycemia exhibited cell hypertrophy, with accompanying increased levels of $p27^{Kip1}$. These findings have been corroborated in other experimental models of diabetic nephropathy, including the type 1 model induced by streptozotocin administration [12] and the type 2 *db/db* mouse model. We have shown that diabetic p21–/– mice are protected from glomerular hypertrophy and the development of progressive renal failure. Diabetic p27–/– mice exhibit only mild mesangial expansion without glomerular hypertrophy, despite an increase in glomerular transforming growth factor-β (TGF-β) [13].

Further corroborating the role of Cdk inhibitors in diabetic nephropathy, Baba et al. [14] demonstrated that in *db/db* mice, recombinant galectin-9, part of a family of carbohydrate-binding proteins with specific binding affinity for β-galactosidase, effectively blocked expression of $p27^{Kip1}$ and $p21^{Cip1}$. Even though the metabolic milieu was not noticeably altered, with similar levels of hyperglycemia and dyslipidemia in the experimental and control groups, the animals treated with recombinant galectin-9 exhibited less glomerular hypertrophy, urinary albumin excretion, and podocyte hypertrophy. Additionally, inhibiting angiotensin II, an important mediator of diabetic nephropathy, has been shown to decrease p27^{Kip1} mRNA and protein expression in glomeruli and podocytes, and this inhibition was associated with prevention of glomerular hypertrophy in diabetic rats [15].

The above studies demonstrate that Cdk inhibitors may underlie the podocyte's tendency to undergo hypertrophy in states of hyperglycemia. Thus, in both diabetic and nondiabetic diseases, although podocyte hypertrophy is initially compensatory in response to podocyte loss, this response is likely maladaptive and contributes to the development of glomerulosclerosis. These events are governed, in part, by specific cell cycle regulatory proteins, especially Cdk inhibitors.

Podocyte Apoptosis

There is an increasing body of literature showing that apoptosis (programmed cell death) of podocytes is a major cause of reduced podocyte number. Although, traditionally, cell cycle regulatory proteins are considered to be involved solely in proliferation, there is mounting evidence that specific cell cycle proteins may also play important roles in the regulation of apoptosis.

Previous studies have shown increased podocyte apoptosis in TGF-β transgenic mice, CD2AP null mice, and in rats exposed to puromycin aminonucleoside. Indeed, several other potent stimuli for podocyte apoptosis have recently been described, including angiotensin II, mechanical stretch, hyperglycemia-induced reactive oxygen species, and cyclosporine A. In vitro studies have shown that during apoptosis, cells typically exit the cell cycle in late G1, although apoptotic cells can exit during any phase of the cell cycle. We have sought to determine what role, if any, specific cell cycle regulatory proteins have in podocyte apoptosis regulation.

Several studies have highlighted Cdk-inhibitor p21^{Cip1} as a potentially important regulator of apoptosis. It is not entirely clear how p21^{Cip1} protects cells against apoptosis, but it has been proposed that either p21^{Cip1}-mediated cell cycle arrest causes resistance to apoptosis or that p21^{Cip1} blocks apoptosis directly. Studies have shown that p21^{Cip1} may interact with pro-apoptotic molecules such as procaspase 3 and apoptosis signal-regulating kinase 1 (ASK1), inhibiting their activities. However, paradoxically, in some cases, p21^{Cip1} can promote apoptosis. Indeed, the function of p21^{Cip1} can vary substantially, based on concentration, cellular context, subcellular localization, and cytotoxic agent.

As the mechanisms of regulation of apoptosis by p21^{Cip1} are poorly understood, Wada et al. [16] recently examined the role of p21^{Cip1} in TGF-β-induced podocyte apoptosis. In wild-type podocytes, upon exposure to TGF-β, a cytokine important in glomerular disease induction and progression and upregulated in a variety of podocyte diseases, including membranous nephropathy, diabetic nephropathy, and FSGS, p21^{Cip1} levels were increased and were associated with an increase in podocyte apoptosis. To examine the role of p21^{Cip1} in podocyte apoptosis, p21^{Cip1}–/– mouse podocytes in culture were exposed to TGF-β. Interestingly, in p21^{Cip1}–/– podocytes, no significant apoptosis was seen when compared to p21^{Cip1}+/+ cells. This finding was associated with an increase in anti-apoptotic Bcl-2 in p21^{Cip1}–/– podocytes, a finding not observed in p21^{Cip1}+/+ cells. Restoration of p21^{Cip1} by an adenoviral transfection prevented the TGF-β-induced increase in Bcl-2 expression, resulting in increased apoptosis. These findings suggest that the upregulation of anti-apoptotic Bcl-2 is suppressed in the presence of the Cdk inhibitor p21^{Cip1} under disease conditions, supporting a critical role for p21^{Cip1} in triggering apoptosis in response to TGF-β-induced podocyte injury [16].

Indeed, upon attempting to unravel the protective effects of glucocorticoids in nephrotic syndrome, beyond nonspecific immunosuppressive or anti-inflammatory effects, Wada et al. [17] recently showed that dexamethasone markedly reduced apoptosis in cultured podocytes following exposure to both puromycin aminonucleoside and TGF-β. This nonimmunological effect of dexamethasone was mediated by a decrease in tumor suppressor p53. Xing et al. [18] have shown that dexamethasone also leads to marked downregulation of p21^{Cip1} expression in cultured human podocytes, further supporting that p21^{Cip1} may play an important role in regulating apoptosis in podocytes.

Expanding our knowledge of cyclins and their regulation of apoptosis in the podocyte, Griffin et al. [19] have recently shown that cyclin I, a nonclassical cyclin most abundantly present in post-mitotic tissues, is expressed exclusively in podocytes in the kidney. Cyclin I –/– mice demonstrated no apparent developmental defects or phenotypic abnormalities under physiological conditions; and cyclin I –/– podocytes in culture showed no

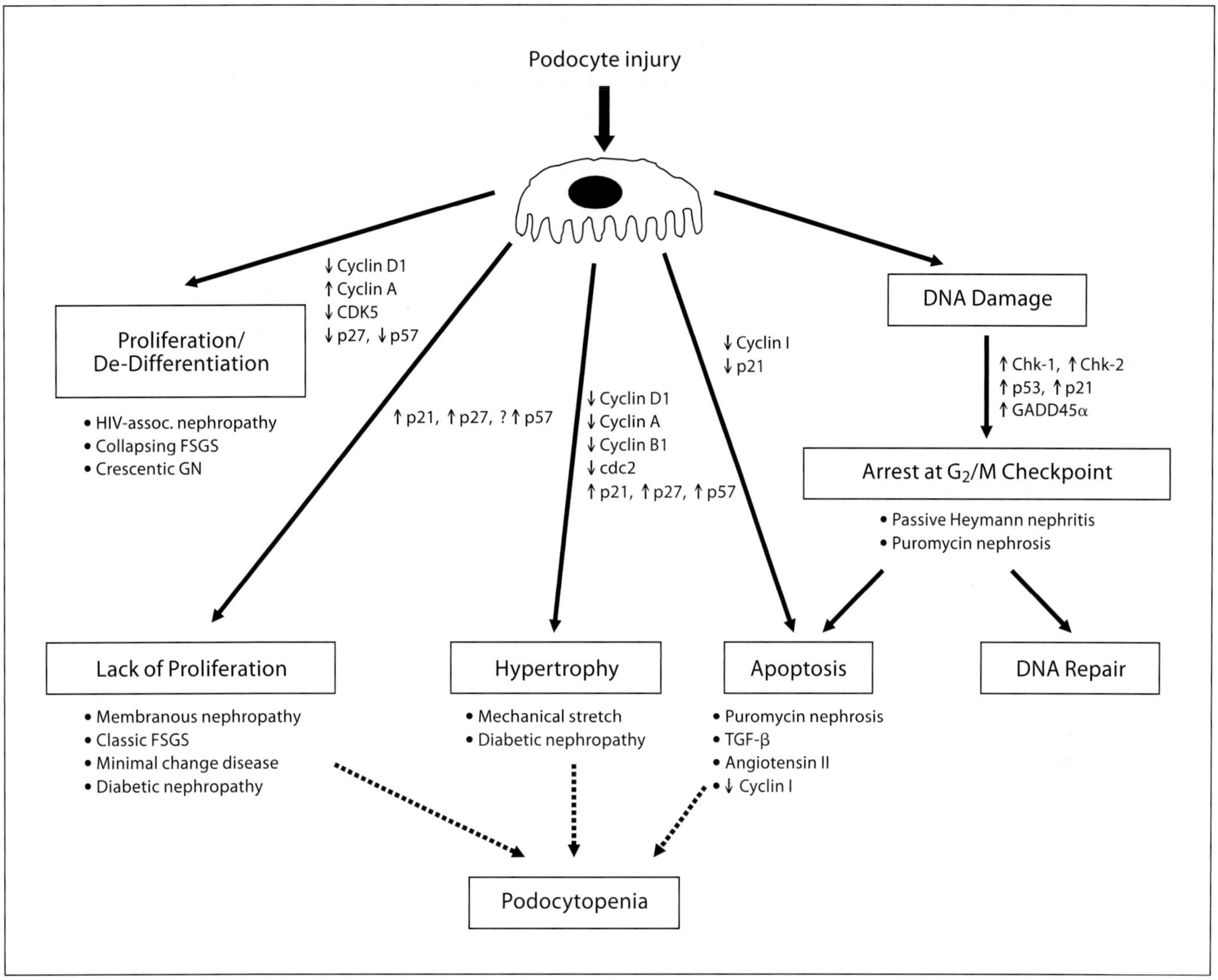

Fig. 1. Cell cycle regulatory proteins in the podocyte's response to injury. Following injury, the normally quiescent and terminally differentiated podocyte can undergo several cell fates, depending on the type of injury. De-differentiation, due to a decrease in Cdk inhibitors, permits cell cycle re-entry, and proliferation occurs when cyclins and Cdks are activated. In contrast, most forms of injury lead to reduction in podocyte number (podocytopenia), due to apoptosis, hypertrophy, lack of proliferation, or the presence of DNA damage. Levels of positive regulators (cyclins and Cdks) and negative regulators (Cdk inhibitors and DNA damage response effectors) may determine which fate the injured cell will follow. Chk-1 = Checkpoint kinase-1; Chk-2 = checkpoint kinase-2; FSGS = focal segmental glomerulosclerosis; GADD45α = growth arrest and DNA damage-inducible-45α; GN = glomerulonephritis; TGF-β = transforming growth factor-β.

detectable defects in cell proliferation. However, upon exposure to apoptotic triggers, including UV-C irradiation, puromycin aminonucleoside, and anti-podocyte antibody, cyclin I –/– podocytes demonstrated earlier onset and increased magnitude of apoptosis than cyclin I +/+ podocytes. Upon stable reconstitution of cyclin I, the apoptotic susceptibility of cyclin I –/– podocytes was reversed. Correlating with the in vitro findings, in the setting of experimental glomerulonephritis induced with a sheep anti-podocyte antibody, cyclin I –/– mice demonstrated worse disease, including increased apoptosis, extracellular matrix expansion, glomerulosclerosis, proteinuria, and renal dysfunction, compared with cyclin I +/+ mice. Interestingly, $p21^{Cip1}$ levels were reduced in cyclin I –/– cells and the re-introduction of cyclin I increased $p21^{Cip1}$ levels, suggesting that a key mechanism by which cyclin I protects podocytes from apoptosis may be via stabilization of $p21^{Cip1}$. Taken together, the above studies underscore that $p21^{Cip1}$ can exhibit both anti-apoptotic and pro-apoptotic functions in podocytes. The factors that determine the precise role of $p21^{Cip1}$ in podocyte apoptosis in response to various disease triggers remain poorly defined and merit further study.

Cell Cycle Proteins and Podocyte Shape

Recently, Cdk5, a nonclassical and novel Cdk that is expressed in certain post-mitotic cells, has been shown to increase in podocytes during differentiation in culture and in developing fetal kidneys. Cdk5 declines in experimental diseases associated with podocyte de-differentiation and proliferation, including antiglomerular antibody disease and HIV-associated nephropathy, implicating a role for Cdk5 in maintaining the mature podocyte phenotype. Indeed, in vitro, inhibiting Cdk5 activity, pharmacologically with roscovitine and with small interfering RNA, significantly altered podocyte shape, with a dramatic reorganization of the actin cytoskeleton. Cdk5 inhibition had no effect on proliferation. Accordingly, a paradigm is developing that Cdk5 is involved in maintaining the podocyte's complex morphology, critical for its highly specialized function [20].

Conclusion

A review of the growing body of literature supporting a critical role for cell cycle regulatory proteins in diseases of the podocyte has been provided (fig. 1). At the level of the cell cycle, both a failure to regenerate lost podocytes and an inappropriate proliferative response can have profound consequences for glomerular structure and function. Indeed, the late changes of increased extracellular matrix and progressive glomerulosclerosis are preceded by, and closely linked to, the various fates that the podocyte can follow in response to injury including proliferation, apoptosis, and hypertrophy, all of which are regulated at the level of the cell cycle. We anticipate that our attempts to understand the mechanisms of cell cycle regulatory proteins in the pathophysiology of diseases of the podocyte will provide exciting new directions for research and ultimately will lead to the development of novel therapeutic strategies to preserve renal function.

Acknowledgements

This work was supported by National Institutes of Health grants to S.J.S. (DK60525, DK56799, DK 51096), by National Institutes of Health grant to C.B.M. (3 RO1 DK051096-06S1), and by the American Diabetes Association. S.J.S. is also an Established Investigator of the American Heart Association.

References

1 Griffin SV, Pichler R, Dittrich M, Durvasula R, Shankland SJ: Cell cycle control in glomerular disease. Springer Semin Immunopathol 2003;24:441–457.

2 Shankland SJ, Eitner F, Hudkins KL, Goodpaster T, D'Agati V, Alpers CE: Differential expression of cyclin-dependent kinase inhibitors in human glomerular disease: role in podocyte proliferation and maturation. Kidney Int 2000;58:674–683.

3 Barisoni L, Mokrzycki M, Sablay L, Nagata M, Yamase H, Mundel P: Podocyte cell cycle regulation and proliferation in collapsing glomerulopathies. Kidney Int 2000;58:137–143.

4 Nelson PJ, Sunamoto M, Husain M, Gelman IH: HIV-1 expression induces cyclin D1 expression and pRb phosphorylation in infected podocytes: cell-cycle mechanisms contributing to the proliferative phenotype in HIV-associated nephropathy. BMC Microbiol 2002;2:26.

5 Moeller MJ, Soofi A, Hartmann I, Le Hir M, Wiggins R, Kriz W, Holzman LB: Podocytes populate cellular crescents in a murine model of inflammatory glomerulonephritis. J Am Soc Nephrol 2004;15:61–67.

6 Shankland SJ, Floege J, Thomas SE, Nangaku M, Hugo C, Pippin J, Henne K, Hockenberry DM, Johnson RJ, Couser WG: Cyclin kinase inhibitors are increased during experimental membranous nephropathy: potential role in limiting glomerular epithelial cell proliferation in vivo. Kidney Int 1997;52:404–413.

7 Kim YG, Alpers CE, Brugarolas J, Johnson RJ, Couser WG, Shankland SJ: The cyclin kinase inhibitor p21CIP1/WAF1 limits glomerular epithelial cell proliferation in experimental glomerulonephritis. Kidney Int 1999;55:2349–2361.

8 Pippin JW, Durvasula R, Petermann A, Hiromura K, Couser WG, Shankland SJ: DNA damage is a novel response to sublytic complement C5b-9-induced injury in podocytes. J Clin Invest 2003;111:877–885.

9 Marshall CB, Pippin JW, Krofft RD, Shankland SJ: Puromycin aminonucleoside induces oxidant-dependent DNA damage in podocytes in vitro and in vivo. Kidney Int 2006;70:1962–1973.

10 Petermann AT, Pippin J, Durvasula R, Pichler R, Hiromura K, Monkawa T, Couser WG, Shankland SJ: Mechanical stretch induces podocyte hypertrophy in vitro. Kidney Int 2005;67:157–166.

11 Hoshi S, Shu Y, Yoshida F, Inagaki T, Sonoda J, Watanabe T, Nomoto K, Nagata M: Podocyte injury promotes progressive nephropathy in Zucker diabetic fatty rats. Lab Invest 2002;82:25–35.

12 Kuan CJ, al-Douahji M, Shankland SJ: The cyclin kinase inhibitor p21WAF1, CIP1 is increased in experimental diabetic nephropathy: potential role in glomerular hypertrophy. J Am Soc Nephrol 1998;9:986–993.

13 Awazu M, Omori S, Ishikura K, Hida M, Fujita H: The lack of cyclin kinase inhibitor p27(Kip1) ameliorates progression of diabetic nephropathy. J Am Soc Nephrol 2003;14:699–708.

14 Baba M, Wada J, Eguchi J, Hashimoto I, Okada T, Yasuhara A, Shikata K, Kanwar YS, Makino H: Galectin-9 inhibits glomerular hypertrophy in db/db diabetic mice via cell-cycle-dependent mechanisms. J Am Soc Nephrol 2005;16:3222–3234.
15 Xu ZG, Yoo TH, Ryu DR, Cheon Park H, Ha SK, Han DS, Adler SG, Natarajan R, Kang SW: Angiotensin II receptor blocker inhibits p27Kip1 expression in glucose-stimulated podocytes and in diabetic glomeruli. Kidney Int 2005;67:944–952.
16 Wada T, Pippin JW, Terada Y, Shankland SJ: The cyclin-dependent kinase inhibitor p21 is required for TGF-beta1-induced podocyte apoptosis. Kidney Int 2005;68:1618–1629.
17 Wada T, Pippin JW, Marshall CB, Griffin SV, Shankland SJ: Dexamethasone prevents podocyte apoptosis induced by puromycin aminonucleoside: role of p53 and Bcl-2-related family proteins. J Am Soc Nephrol 2005;16:2615–2625.
18 Xing CY, Saleem MA, Coward RJ, Ni L, Witherden IR, Mathieson PW: Direct effects of dexamethasone on human podocytes. Kidney Int 2006;70:1038–1045.
19 Griffin SV, Olivier JP, Pippin JW, Roberts JM, Shankland SJ: Cyclin I protects podocytes from apoptosis. J Biol Chem 2006;281:28048–28057.
20 Griffin SV, Hiromura K, Pippin J, Petermann AT, Blonski MJ, Krofft R, Takahashi S, Kulkarni AB, Shankland SJ: Cyclin-dependent kinase 5 is a regulator of podocyte differentiation, proliferation, and morphology. Am J Pathol 2004;165:1175–1185.

nephron
Experimental
Nephrology

Nephron Exp Nephrol 2007;106:e60–e66
DOI: 10.1159/000101794

Published online: June 6, 2007

Role of Transcription Factors in Podocytes

Anne Rascle Hani Suleiman Tanja Neumann Ralph Witzgall

Institute for Molecular and Cellular Anatomy, University of Regensburg, Regensburg, Germany

Key Words
Wilm's tumor gene 1 · Slit diaphragm · Foot processes

Abstract
Despite a wealth of information on structural proteins, comparatively little is known on the transcriptional regulation of podocyte structure and function. In this review we will highlight those transcription factors which, by gene inactivation or classical transgenic experiments, have been shown to be essential for podocytes or probably will turn out to be so. The tumor suppressor protein WT1 is not only indispensable for the initial stages of kidney development, but also very likely maintains the integrity of the fully differentiated podocyte. In the kidney, the LIM homeodomain transcription factor LMX1B is specifically synthesized in podocytes, and mutations in *LMX1B* lead to nail-patella syndrome and the associated nephropathy. Other transcription factors such as hypoxia-inducible factors and PAX2 are likely to play a role in podocytes, whereas the significance of others, e.g. of POD1 and CITED2, is more speculative at this point.

Introduction

The essential function of podocytes was irrevocably brought to nephrologists' attention with the identification of the *NPHS1* gene [1], the gene mutated in patients suffering from congenital nephrotic syndrome of the Finnish type. From then on the podocyte has not left center stage when it comes to the investigation of glomerular development and diseases. During development, podocytes differentiate from a cuboidal into an octopus-shaped cell with many large and small tentacles. Primary processes (the large tentacles) extend from the cell body and elaborate many fine secondary processes, the foot processes. By a mechanism so far shrouded in mystery, foot processes emanating from two different primary processes (but not from the same primary process!) are connected by a proteinaceous bridge called the slit diaphragm, one component of the glomerular filtration barrier. In the absence of the protein product of the *Nphs1* gene, nephrin, no slit diaphragms form [2, 3]. Both figuratively and literally speaking, nephrin has been the nucleus around which the slit diaphragm was built. Furthermore, the podocytes, together with glomerular endothelial cells, are also responsible for the synthesis of the glomerular basement membrane. Despite this increasing wealth of information on structural proteins produced by

KARGER

Fax +41 61 306 12 34
E-Mail karger@karger.ch
www.karger.com

1660–2129/07/1062–0060$23.50/0

Accessible online at:
www.karger.com/nee

Ralph Witzgall
University of Regensburg, Institute for Molecular and Cellular Anatomy
Universitätsstrasse 31, DE–93053 Regensburg (Germany)
Tel. +49 941 943 2820, Fax +49 941 943 2868
E-Mail ralph.witzgall@vkl.uni-regensburg.de

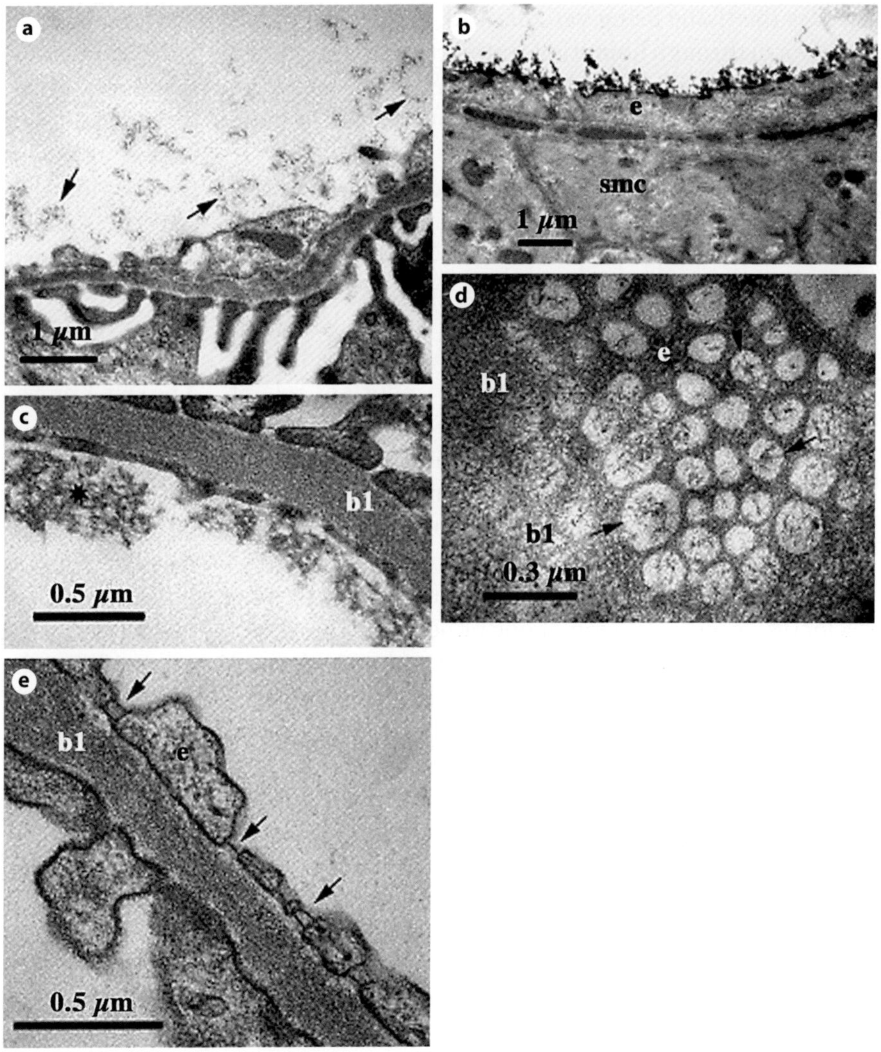

Fig. 1. Ultrastructural demonstration of the glomerular endothelial cell surface layer (ESL) and glycocalyx. Tissue was fixed in oxygen-rich fluorocarbon-based perfusion buffer followed by tannic acid treatment (**a**, **b**), cupromeronic blue (Cmb) cytochemistry for proteoglycans (**c**, **d**), or osmication in the presence of potassium ferricyanide (**d**). **a** Arrows point to a loose network of fibrillar strands extending >1 μm from the endothelial cell surface. **b** An arteriole subjected to the same treatment as in **a**. e = Endothelium; smc = smooth muscle cell. **c** Glomerular endothelium covered by material that forms patches (asterisk). Shrinkage of endothelial cells is due to the method of preparation. bl = Basal lamina. **d** En face image of the fenestrated portion of a glomerular endothelial cell (e). bl = Material covering the basal lamina. The proteoglycans stained by Cmb form a loose network in the fenestrated region of the endothelial cell (arrows). **e** Single and double diaphragms bridge glomerular endothelial cell fenestrae, shown by arrows. Reproduced from Hjalmarsson et al. [15], with permission.

lar endothelial cells are covered by a 10- to 20-nm glycocalyx that is an integral part of their plasma membranes and fenestrae. The ESL forms an additional loosely associated layer. Current functional and structural studies strongly suggest that the glycocalyx and ESL contribute substantially to the barrier function of the glomerular endothelium towards macromolecules.

Mediators of Glomerular Endothelial Cell Development and Differentiation

Glomerular capillaries develop through local vasculogenesis, which involves migration of individual angioblasts into the capillary cleft of developing nephrons, followed by their assembly into capillaries. Neutralizing antibodies to VEGF-A [18] or transforming growth factor-β1 (TGF-β1) [19] markedly inhibit angioblast migration into nascent glomeruli, indicating that both VEGF-A and TGF-β1 are critically important for glomerular capillary formation. Invasion of the vascular cleft by angioblasts furthermore is guided by neuropilin, the receptor for semaphorin III which also binds the 165-kDa isoform of VEGF-A, and a role for podocyte ephrin-B2 in guiding angioblasts into the vascular clefts of developing nephrons has been postulated [20].

Upon arrival in the glomerular capillary cleft, angioblasts first form precapillary cords without recognizable lumens [19, 21–23]. Formation of capillary lumens then requires removal of superfluous endothelial cells by TGF-β1-dependent apoptosis. Hence, unlike the process of angiogenesis in which a sprouting vessel containing a lumen invades the tissue being vascularized, glomerular capillaries form through homotypic interaction of angioblasts to form a precapillary cord. The lumen is then shaped as a distinct event requiring apoptosis of a subset of these endothelial cells.

The observation that capillaries near epithelial cells expressing high levels of VEGF are generally fenestrated led Breier et al. [24] to propose that VEGF is necessary for the development of fenestrae in ECs. Among the various tissues they examined, they found the highest VEGF expression in mature glomerular podocytes. Indeed, most of the current evidence points to an important role of podocytes in directing glomerular EC development and differentiation. For instance, deletion of the podocyte-specific transcription factor LMX1B in mice, which prevents podocyte differentiation and podocyte VEGF-A expression, is associated with the formation of glomerular endothelium that fails to develop fenestrae [25]. Using podoctye-specific deletion of one or both VEGFA alleles in mice, Eremina et al. [26] proved that podocyte-derived VEGF is critical both for the initial ingrowth of glomerular ECs into glomeruli and for their fenestration.

The site-specific differentiation of glomerular ECs also depends on the composition of its extracellular matrix and on the pleotropic mediator TGF-β1. Inhibition of TGF-β1 in developing rat kidney prevented the formation of a flattened, fenestrated glomerular endothelium [19]. Mutations in laminin β3 gene (LAMB3) that cause the human disorder 'Herlitz junctional epidermolysis bullosa' are associated with profound congenital nephrotic syndrome and a lack of glomerular EC fenestrae [27]. In mice, deletion of the laminin α3 gene *(Lama3)* produces a similar blistering skin disease and a complete absence of glomerular EC fenestrae [28]. While it is safe to conclude that formation of glomerular EC fenestrae is directed by the local microenvironment, with VEGF-A, TGF-β1 and laminin subtype playing important roles, the precise mechanisms that regulate fenestrae formation are far from understood.

Loss of the Differentiated Glomerular Endothelial Cell Phenotype in Pre-Eclampsia

The characteristic glomerular lesion of pre-eclampsia comprises the loss of glomerular endothelial cell fenestrae, glomerular endothelial cell swelling and proteinuria, often in the nephrotic range. The ultrastructural appearance of the glomerular endothelium in pre-eclampsia is referred to as glomerular 'endotheliosis'. The glomerular abnormalities of pre-eclampsia are recapitulated in mice in which one allele of VEGF-A is deleted from podocytes [26]. In recent years, Karumanchi and coworkers [29] have been successful in defining the nature of at least two soluble mediators that can account for the glomerular phenotype observed in pre-eclampsia. The soluble form of VEGF receptor-1 (sVEGFR-1, sFlt-1) which neutralizes the angiogenic mediators VEGF and placental growth factor (PlGF) is found at much higher concentrations in the plasma of patients with pre-eclampsia than in women with normal pregnancies [30]. The sFlt-1 prevents the interaction of VEGF-A with full-length transmembrane VEGF receptors. Since there is good evidence that VEGF is required for glomerular endothelial cell fenestrae formation, it follows that inhibition of VEGF-A by sFlt-1 could result in the loss of glomerular endothelial cell fenestrae. Indeed, when neutralizing VEGF antibodies or sFlt-1 are infused in mature mice, rapid loss of endothelial cell fenestrae, endothelial cell swelling, and proteinuria are observed [31]. Similarly, in rats sFlt-1 also produces the glomerular endothelial cell lesion found in pre-eclampsia [32]. More recent work, also from the Karumanchi group [33], has shown that the plasma concentration of second soluble receptor, sEndoglin is also elevated in patients with pre-eclampsia. Endoglin is an endothelial cell-restricted type III receptor for TGF-β1 required for activation of the ALK1 signaling cascade in endothelial cells, and its deletion in mice is embryonic lethal due to a defect in vasculogenesis [34]. Furthermore, TGF-β1 is required for glomerular endothelial fenestrae formation [19]. It can therefore be concluded that increased release of sFlt-1 and sEndoglin by the placenta account for the glomerular endothelial cell abnormalities observed in pre-eclampsia. Given the role of VEGF in regulating glomerular endothelial cell differentiation and fenestrae formation, it is also not surprising that the use of neutralizing VEGF antibodies in the treatment of various malignancies is associated with glomerular injury, most notably proteinuria [35].

Taken together, the developmental studies and studies in mice with podocyte-specific deletion of VEGF-A prove that the final differentiation of glomerular endothelial cells is podocyte-dependent, and that podocyte-derived VEGF-A is a necessary mediator of glomerular endothelial cell differentiation. The rapid reversal of endothelial cell fenestration by neutralizing VEGF-A antibodies suggests that the formation of fenestrae is a very dynamic process.

cal observations, mainly on patients with type 1 diabetes, indicate that there is a latent period of several years after the onset of diabetes which is characterized by nephromegaly and increased glomerular filtration rate without other clinical abnormalities. Thereafter, and in those ~30% of diabetic patients genetically destined to develop overt nephropathy, microalbuminuria develops as the first clinical manifestation of the disease. Microalbuminuria can progress, if untreated, to overt proteinuria, sometimes in the nephrotic range, and ultimately renal insufficiency. Although cases have been described of end-stage renal disease being reached in diabetes without significant proteinuria, the vast majority of patients have proteinuria before a decrease in renal function occurs. Since proteinuria has important pathophysiologic and clinical implications, the present manuscript will focus on the cellular and molecular mechanisms of proteinuria in diabetes mellitus.

Hemodynamic Changes

Glomerular hemodynamic changes including hyperperfusion and hyperfiltration are found very early in the disease process [1]. These hemodynamic changes facilitate the development of albumin leakage from the glomerular capillary compartment into Bowman's space. The increase in glomerular plasma flow and the elevated glomerular transcapillary hydrostatic pressure result from a decrease in both afferent and efferent arteriolar resistances wherein the afferent arteriole is more dilated than the efferent arteriole [1]. Formally, these changes are ascribed to defective autoregulation because in the normal situation an increase in perfusion pressure would result in preglomerular vasoconstriction so as to maintain a constant glomerular filtration rate. Many diverse factors including prostanoids, nitric oxide (NO), atrial natriuretic factor, growth hormone, glucagon, insulin, angiotensin II (ANG II), and others have been implicated as agents causing hyperperfusion and hyperfiltration [for review, see 2]. Elevated intraglomerular pressure has been linked to the development of mesangial cell matrix overproduction and podocyte injury with thickening of the glomerular basement membrane (GBM) and the eventual development of glomerulosclerosis. Other factors that are induced by the diabetic environment which could influence glomerular hemodynamics are vascular endothelial growth factor-A (VEGF-A), perhaps via production of NO, and even cytokines such as transforming growth factor-β1 (TGF-β1) which may mediate hyperfiltration by arteriolar vasodilatation via inhibiting calcium transients [2, 3]. Furthermore, TGF-β1 increases NO production in early diabetes, presumably by upregulation of endothelial NO synthase (eNOS) mRNA expression and enhancing arginine resynthesis [3]. Thus, TGF-β1 could clearly play a role in diabetic vascular dysfunction and may mediate some of the hemodynamic changes associated with early diabetic nephropathy [3]. On the other hand, shear stress and mechanical strain, resulting from altered glomerular hemodynamics, induce autocrine and/or paracrine release of cytokines and growth factors. For instance, exposing mesangial cells, glomerular endothelial cells, or podocytes in culture to stretching or shear stress can activate certain signal transduction systems, growth responses, cytokine synthesis (e.g. ANG II, TGF-β), and increased production of extracellular matrix proteins [2]. Such studies suggest that the local activation of cytokines and growth factors can mechanistically link the local hemodynamic stress to the structural changes of diabetic nephropathy.

As an alternative explanation, glomerular hyperfiltration in diabetic nephropathy has been linked to a primary abnormality in tubular sodium reabsorption [4]. Within the context of tubuloglomerular feedback, this explanation suggests that an increased reabsorption of sodium chloride in the proximal tubule or loop of Henle leads to an increase in glomerular filtration rate by an intact macula densa mechanism [4]. Diabetes-induced hypertrophy of tubules which mediates stimulated sodium chloride reabsorption could be pivotal in this process, thus linking again renal structural changes to the hemodynamic adaptation of diabetic kidney disease [4]. Irrespective of what causes the intraglomerular hemodynamic changes, the development and progression of albuminuria are accelerated by the hemodynamic stress.

Structural Abnormalities

One of the established morphological changes of diabetic glomerulopathy is the expansion of the mesangial area [5]. This is caused by an increase in extracellular matrix deposition (mostly collagen IV and fibronectin) and mesangial cell hypertrophy [5]. Thickening of the GBM occurs early and is detectable in the first 1–2 years after the onset of diabetes. Both increased extracellular matrix protein synthesis and impaired protein degradation contribute to the thickening. In addition to an increase in collagen type IV deposition in the GBM, the expression of heparan sulfate proteoglycans and the extent of sulfation decreases. Therefore, not only is there a quantitative increase in the GBM but there is a qualitative change that may promote macromolecular permeability. As far as collagen IV is concerned, there is increased deposition of α1(IV) and α2(IV) chains in the mesangial area in diabetic nephropathy, whereas α3(IV) and α4(IV) levels are increased in the GBM [6, 7]. Late in the course of glomerulosclerosis there is de novo deposition of fibrillar collagen types I and III in the mesangial area.

Podocyte Injury in Human Diabetes

Although mesangial cell pathophysiology has long been considered at the heart of the development of diabetic nephropathy ('the mesangiocentric view'), more recent evidence suggests that the podocyte is involved early in the course of kidney disease in both type 1 and type 2 diabetes mellitus. While earlier histomorphometric studies in patients with type 1 diabetes and advanced nephropathy with frank proteinuria revealed a significant increase in the width of podocyte foot processes, more recent investigations in microalbuminuric type 1 diabetic patients also described an increase in foot process width [8]. A cross-sectional analysis of renal biopsies from the European Study for the Prevention of Renal Disease in Type 1 Diabetes (ESPRIT) in patients with increased urinary albumin excretion, but with normotension, showed no significant reduction in podocyte number compared with nondiabetic controls. However, longitudinal studies clearly demonstrated a reduction in podocyte number during follow-up that closely correlated with proteinuria. Renal biopsies from Pima Indians with type 2 diabetes showed a broadening in podocyte foot processes and a concomitant reduction in the number of podocytes per glomerulus [8].

In order to detect earlier functional abnormalities in podocytes, more recent studies have focused on expression of specific podocyte-associated proteins in diabetic patients. It has been found that nephrin excretion in the urine was present in 30% of normoalbuminuric, 17% of microalbuminuric, and 28% of macroalbuminuric type 1 diabetic patients whereas none of the healthy subjects had nephrinuria. There is also downregulation of nephrin protein expression in the slit diaphragm in diabetic patients compared with nondiabetic controls [9]. This decrease in nephrin staining correlates inversely with foot process width, and underscores the important role of nephrin in the integrity of the podocyte. In contrast, expression of CD2AP, another podocyte-specific protein, is not reduced in podocytes from diabetic patients, suggesting that the observed reduction in nephrin is not due to general loss or major injury of podocytes.

Podocyte Dysfunction in Diabetes as Revealed in Experimental Models

Animal experiments allow longitudinal study of the development and progression of diabetic nephropathy and the potential for pharmacological interventions. Serial investigations on kidneys from Zucker *fa/fa* rats, a model of type 2 diabetes but with segmental glomerulosclerosis, revealed that nephropathy started with damage to podocytes, including foot process effacement, and cytoplasmic accumulation of lipid droplets. Combined treatment with an ACE inhibitor and an AT1-receptor antagonist was more effective in controlling renal damage, including podocyte injury and nephrin suppression, in obese Zucker rats than with either monotherapy alone. Thus, a local increase in ANG II is associated with podocyte injury, including downregulation of nephrin expression.

The detailed molecular mechanisms for ANG II-induced nephrin suppression are still incompletely understood. Podocytes express AT1- and probably AT2-receptors [10]. ANG II transactivates the epidermal growth factor receptor in podocytes by activation of AT1- and AT2-receptors [11]. This is caused by ANG II-induced processing of heparin-binding-epidermal growth factor (HB-EGF) from its precursor [11]. Moreover, ANG II-stimulated release of HB-EGF activates in a paracrine fashion MAP kinase activity in adjacent cells [11]. Transgenic rats with targeted overexpression of the AT1-receptor in podocytes show pseudocysts in podocytes, followed by foot process effacement and local detachments [12]. These changes subsequently progress to focal segmental glomerulosclerosis. It has been previously shown that high ambient glucose concentration induces ANG II formation in podocytes through upregulation of angiotensinogen expression [12]. Furthermore, proteinuria and protein passage through the ultrafiltration barrier locally activate ANG II formation in podocytes. Finally, mechanical stretch could increase ANG II generation in podocytes. Some evidence indicates that an increase in intracellular reactive oxygen species (ROS) is important as a signal transmitter of this increase in angiotensinogen expression [12]. Interestingly, ANG II formation as a consequence of mechanical stress appears independent of ACE. In this regard it has been demonstrated that chymase, an ANG II-forming enzyme not inhibited by ACE inhibitors, is upregulated in glomeruli of patients with nephropathy due to type 2 diabetes. This suggests that local glomerular formation of ANG II in the diabetic state may be partly independent of ACE and is not abolished by ACE inhibition alone. As a consequence, an increase in local ANG II in the podocyte microenvironment suppresses nephrin expression, and thereby enhances ultrafiltration of proteins that further contributes to podocyte damage [13]. Interestingly, recent data suggest that treatment with the glitazones prevents the downregulation of nephrin in the diabetic state [14].

A loss of nephrin and proteoglycan expression may not be the only pathological change fostering proteinuria in diabetic nephropathy. VEGF is an angiogenic factor that markedly increases vascular permeability [12]. The expression of VEGF is pronounced in podocytes, and experimental diabetes increases VEGF mRNA and protein expression. Exposure of cultured podocytes to high glucose concentrations also upregulates VEGF expression [15]. Also, ANG II appears to be an important mediator of the diabetes-associated increase in VEGF synthesis in podocytes because an AT1-receptor antagonist attenuates these changes [15]. In addition, ANG II directly stimulates VEGF expression in cultured podocytes even under normal glucose conditions [15]. ANG II-mediated VEGF production in the kidney is not only mediated by AT1- but also AT2-receptors [12]. In fact, only an ACE inhibitor, but not an AT1-receptor antagonist, attenuated enhanced renal VEGF protein concentrations in streptozotocin-induced diabetic rats [12]. An important stimulatory transcription factor for VEGF expression is hypoxia-inducible factor-1 (HIF-1). We have recently shown that ANG II stimulates HIF-1α expression by a posttranscriptional mechanism via AT2-receptors [12]. This increase is caused by downregulation of SM-20/PHD3, the enzyme responsible for hydroxylation of HIF-1α initiating degradation [12].

Rincon-Choles et al. [16] recently discovered that expression of ZO-1, a junctional protein involved in the assembly of tight junctions and expressed at the podocyte slit diaphragm, is reduced in animal models of types 1 and 2 diabetes. Exposure of podocytes to high glucose in vitro also resulted in a decrease in ZO-1 expression [16]. This effect can contribute to alterations in podocyte structure and function that enhance proteinuria.

Proteinuria itself could amplify podocyte injury further by a positive feedback cycle. In rats with reduced renal mass, plasma protein accumulation in podocytes,

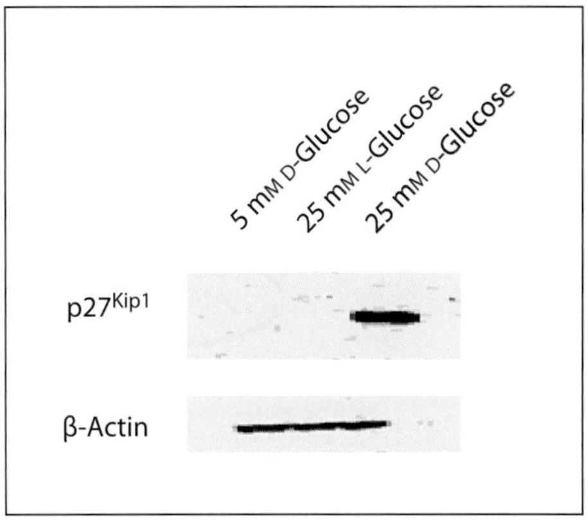

Fig. 1. Induction of p27^{Kip1} by high glucose in cultured podocytes. This Western blot shows induction of the cell cycle inhibitor p27^{Kip1} by 25 mM D-glucose compared with normal glucose (5 mM D-glucose). This effect is independent of osmolarity because 25 mM L-glucose, which is not taken up by cells, has no effect. Upregulation of p27^{Kip1} in podocytes sensitizes these cells for hypertrophy and subsequent apoptosis.

which is a consequence of enhanced transcapillary passage, induces the profibrotic cytokine TGF-β1 [17]. Both albumin and high ambient glucose stimulate TGF-β1 expression in cultured podocytes [12]. An increase in TGF-β1 can stimulate extracellular matrix synthesis by the podocyte itself in an autocrine manner or may reach mesangial cells to induce a sclerosing phenotype in these cells. Advanced glycation end-products (AGEs) are also involved in induction of TGF-β1 in podocytes. Expression of RAGE, the receptor for AGEs, is increased in podocytes in diabetic *db/db* mice [12]. Activation of RAGE leads to increased VEGF and TGF-β1 expression [12]. We recently reported that blockade of VEGF signaling by a receptor kinase inhibitor ameliorates diabetic albuminuria in *db/db* mice [18]. These in vivo data support a link between long-term metabolic events in diabetes such as AGEs to increased expression of VEGF and TGF-β1 and eventually to podocyte pathology that manifests in proteinuria.

An increase in TGF-β1 may have additional effects besides stimulation of extracellular matrix. In podocytes, high levels of TGF-β1 induce apoptosis through Smad7 by inhibiting nuclear translocation of the cell survival factor NF-κB [19]. TGF-β1-mediated activation of Smad7 appears to be specific for podocytes and is not found in mesangial cells, as assessed in a limited biopsy series of diabetic nephropathy patients [19]. Since podocytes are terminally differentiated cells that are unable to undergo cell division because of upregulation of cell cycle inhibi-

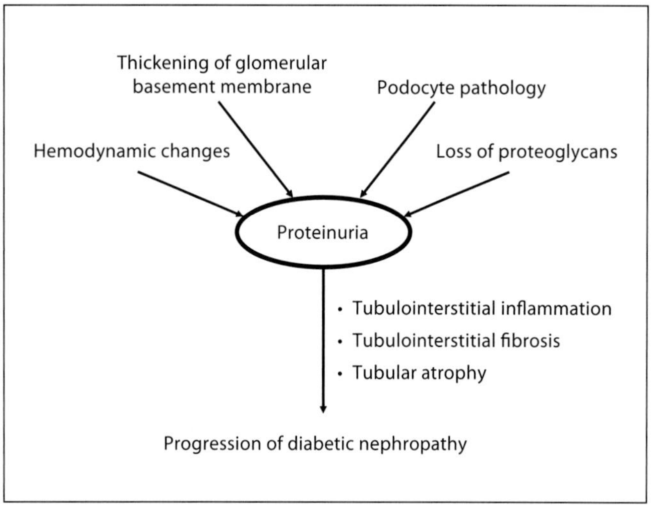

Fig. 2. Overview of mechanisms leading to proteinuria in diabetic nephropathy. Although hemodynamic factors (e.g. glomerular hypertension and hyperfiltration), thickening of the GBM, and loss of negatively charged proteoglycans are important factors for the development of proteinuria, recent research has focused on the primary role of podocyte pathology in this process (for details see text). Proteinuria induces in tubular cells a cascade of events that leads to interstitial inflammation/fibrosis and tubular atrophy.

tory proteins p57 and $p27^{Kip1}$, apoptotic loss of cells cannot be replaced [20]. High glucose additionally stimulates the expression of $p27^{Kip1}$ resulting in cell cycle arrest before successful replication (fig. 1). We found recently that AGEs also stimulate the expression of $p27^{Kip1}$ in podocytes, likely enhancing the maladaptive phenotype leading to apoptosis [Wolf et al., unpubl. observations]. $p27^{Kip1}$ knockout mice are partially protected from the development of diabetic nephropathy, and morphometric analysis indicates that there is less podocyte injury in diabetic $p27^{Kip1}$ knockout mice compared with diabetic wild-type animals, suggesting an important role of $p27^{Kip1}$ in podocyte injury [20].

Apoptosis of podocytes can further enhance proteinuria. Susztak et al. [21] discovered that glucose-mediated ROS generation, which is a common upstream initiator of many pathophysiological abnormalities in diabetes mellitus, is responsible for podocyte depletion early on in experimental diabetic nephropathy. ANG II contributes to renal ROS formation through activation of NADPH oxidase. Moreover, depletion of podocytes allows the denuded GBM to come into close contact with Bowman's capsule, resulting in synechiae formation, the first step in the development of glomerulosclerosis.

Conclusion

Proteinuria, an important contributor to progression of diabetic nephropathy, is caused by several factors. On a molecular level, metabolic pathways activated by hyperglycemia as well as proteins altered by high blood glucose such as Amadori-albumin and AGEs are key players in the development of diabetic nephropathy (fig. 2). Oxidative stress is an early event in the development of diabetic complications. A variety of growth factors and cytokines are then induced through complex signal transduction pathways. The cytokine TGF-β1 has emerged as an important downstream mediator for the development of renal hypertrophy and the accumulation of extracellular matrix components. Besides changes in the GBM with thickening and loss of negatively charged sites that contribute to increase flux of proteins through the glomerular ultrafiltration barrier, recent research has focused on podocyte pathology as a major factor causing proteinuria in diabetic nephropathy. Podocytes are a victim of pathophysiological events associated with diabetic nephropathy and are also targets of the effects of high glucose, AGEs, mechanical stretch, and proteinuria itself. A reduction in nephrin protein at the slit diaphragm is associated with podocyte foot process effacement. Proteinuria itself can then induce tubular atrophy and interstitial fibrosis. Podocyte injury is thus a culprit for progression of diabetic nephropathy. Many of these events are mediated by ANG II whose local concentration could be stimulated by high glucose, mechanical stretch, and proteinuria. ANG II, in turn, stimulates VEGF and suppresses nephrin expression events that can aggravate the proteinuria, and also induces TGF-β1 expression leading ultimately to an increase in podocyte apoptosis and fostering the development of glomerulosclerosis.

Acknowledgements

G.W. is supported by a grant from the Deutsche Forschungsgemeinschaft. F.N.Z. is supported by grants from the American Diabetes Association and the National Institutes of Health (DK44513 and DK54608).

References

1 Hostetter TH: Hyperfiltration and glomerulosclerosis. Semin Nephrol 2003;23:194–199.
2 Chen S, Wolf G, Ziyadeh FN: The renin-angiotensin system in diabetic nephropathy. Contrib Nephrol. Basel, Karger, 2001, vol 135, pp 212–221.
3 Tsuchida K, Cronin B, Sharma K: Novel aspects of transforming growth factor-beta in diabetic kidney disease. Nephron 2002;92: 7–21.
4 Thomson SC, Vallon V, Blantz RC: Kidney function in early diabetes: the tubular hypothesis of glomerular filtration. Am J Physiol 2004;286:F8–F15.
5 Wolf G, Ziyadeh FN: Molecular mechanisms of diabetic hypertrophy. Kidney Int 1999;56: 393–405.
6 Kalluri R: Proteinuria with and without glomerular podocyte effacement. J Am Soc Nephrol 2006;17:2383–2389.
7 Zeisberg M, Ericksen MB, Hamano Y, Neilson EG, Ziyadeh F, Kalluri R: Differential expression of type IV collagen isoforms in rat glomerular endothelial and mesangial cells. Biochem Biophys Res Commun 2002;295: 401–407.
8 Pagtalunan ME, Miller PL, Jumping-Eagle S, Nelson RG, Myers BD, Rennke HG, Coplon NS, Sun L, Meyer TW: Podocyte loss and progressive glomerular injury in type II diabetes. J Clin Invest 1997;99:342–348.
9 Doublier S, Salvidio G, Lupia E, Ruotsalainen V, Verzola D, Deferrari G, Camussi G: Nephrin expression is reduced in human diabetic nephropathy. Evidence for a distinct role for glycated albumin and angiotensin II. Diabetes 2003;52:1023–1030.
10 Liebau MC, Lang D, Bohm J, Endlich N, Bek MJ, Witherden I, Mathieson PW, Saleem MA, Pavenstadt H, Fischer KG: Functional expression of the renin-angiotensin system in human podocytes. Am J Physiol 2006;290: F710–F719.
11 Flannery PJ, Spurney RF: Transactivation of the epidermal growth factor receptor by angiotensin II in glomerular podocytes. Nephron Exp Nephrol 2006;103:e109–e118.
12 Wolf G, Chen S, Ziyadeh FN: From the periphery of the glomerular capillary wall toward the center of disease: podocyte injury comes of age in diabetic nephropathy. Diabetes 2005;54:1626–1634.
13 Gruden G, Perin PC, Camussi G: Insight on the pathogenesis of diabetic nephropathy from the study of podocyte and mesangial cell biology. Curr Diab Rev 2005;1:27–40.
14 Benigni A, Zoja C, Tomasoni S, Campana M, Corna D, Zanchi C, Gagliardini E, Garofano E, Rottoli D, Ito T, Remuzzi G: Transcriptional regulation of nephrin gene by peroxisome proliferator-activated receptor-gamma agonist: molecular mechanism of the antiproteinuric effect of pioglitazone. J Am Soc Nephrol 2006;17:1624–1632.
15 Chen S, Lee JS, Iglesias-de la Cruz MC, Wang A, Izquierdo-Lahuerta A, Gandhi NK, Danesh FR, Wolf G, Ziyadeh FN: Angiotensin II stimulates alpha3(IV) collagen production in mouse podocytes via TGF-beta and VEGF signalling: implications for diabetic glomerulopathy. Nephrol Dial Transplant 2005;20:1320–1328.
16 Rincon-Choles H, Vasylyeva TL, Pergola PE, Bhandari B, Bhandari K, Zhang JH, Wang W, Gorin Y, Barnes JL, Abboud HE: ZO-1 expression and phosphorylation in diabetic nephropathy. Diabetes 2006;55:894–900.
17 Abbate M, Zoja C, Morigi M, Rottoli D, Angioletti S, Tomasoni S, Zanchi C, Longaretti L, Donadelli R, Remuzzi G: Transforming growth factor-beta1 is up-regulated by podocytes in response to excess intraglomerular passage of proteins: a central pathway in progressive glomerulosclerosis. Am J Pathol 2002;161:2179–2193.
18 Sung SH, Ziyadeh FN, Wang A, Pyagay PE, Kanwar YS, Chen S: Blockade of vascular endothelial growth factor signaling ameliorates diabetic albuminuria in mice. J Am Soc Nephrol 2006;17:3093–3104.
19 Schiffer M, Mundel P, Shaw AS, Böttinger EP: A novel role for the adaptor molecule CD2-associated protein in transforming growth factor-β-induced apoptosis. J Biol Chem 2004;279:37004–37012.
20 Shankland SJ: The podocyte's response to injury: role in proteinuria and glomerulosclerosis. Kidney Int 2006;69:2131–2147.
21 Susztak K, Raff AC, Schiffer M, Böttinger EP: Glucose-induced reactive oxygen species cause apoptosis of podocytes and podocyte depletion at the onset of diabetic nephropathy. Diabetes 2006;55:225–233.

nephron
Physiology

Nephron Physiol 2007;106:p32–p37
DOI: 10.1159/000101798

Published online: June 6, 2007

Role of the VEGF-A Signaling Pathway in the Glomerulus: Evidence for Crosstalk between Components of the Glomerular Filtration Barrier

Vera Eremina[a] Hans J. Baelde[a, d] Susan E. Quaggin[a–c]

[a]The Samuel Lunenfeld Research Institute, University of Toronto, [b]Division of Nephrology, St. Michael's Hospital, and [c]Institute of Medical Science and Department of Physiology, University of Toronto, Toronto, Ont., Canada; [d]Department of Pathology, Leiden University Medical Center, Leiden, The Netherlands

Key Words
VEGF-A · Podocyte · Endothelium · Glomerulus · Glomerular filtration barrier

Abstract
Background/Aims: Vascular endothelial growth factor is a major regulator of angiogenesis and vascular permeability [Carmeliet et al.: Nature 1996;380:435–439]. The podocyte, the outermost layer of the glomerular filtration barrier, produces large amounts of VEGF-A. The observation that levels of VEGF-A are altered in glomerular diseases, the identification of a link between pre-eclampsia and elevated levels of a circulating soluble VEGF receptor, and the entry of anti-VEGF therapies into the clinical arena have generated intense interest in the functional role of VEGF-A in the glomerulus. ***Methods:*** A variety of studies have been performed to address the role of VEGF-A signaling in the glomerulus. These include descriptions of expression patterns in human renal biopsies, cell culture studies to dissect paracrine versus autocrine signaling roles, and manipulation of VEGF-A expression in animal models using pharmacologic, biologic or genetic approaches. ***Results:*** Exquisite dosage sensitivity to VEGF-A exists in the developing glomerulus as small reductions in the expression of VEGF-A lead to profound changes in glomerular structure and function in mice. The use of VEGF inhibitors is associated with damage to the glomerular endothelium in animal models and proteinuria in patients, suggesting that local VEGF-A production is also required for maintenance of this specialized vascular bed. ***Conclusions:*** Tight regulation of VEGF-A signaling is required for development and maintenance of the glomerular filtration barrier (GFB) and emphasizes the role of podocyte-endothelial crosstalk in the glomerulus. The relative contributions of various VEGF-A isoforms, the role of autocrine signaling in vivo and identification of factors and mechanisms that regulate constitutive expression, storage and delivery of VEGF-A in the glomerulus are still under investigation.

Role of VEGF-A in Glomerular Development

Vascular endothelial growth factor (VEGF or VEGF-A) belongs to a family of growth factors that include VEGF-B, -C, -D, and -E, placental growth factor (PlGF) and platelet-derived growth factor (PDGF). In all endothelial beds, VEGF-A is a major regulator of angiogenesis and vascular permeability [1–4].

During fetal development, one of the cell types producing the largest amounts of VEGF-A is the podocyte. Unlike many other tissues or cells, podocytes continue to

KARGER
Fax +41 61 306 12 34
E-Mail karger@karger.ch
www.karger.com

1660–2137/07/1062–0032$23.50/0

Accessible online at:
www.karger.com/nep

Susan E. Quaggin
The Samuel Lunenfeld Research Institute
600 University Avenue, University of Toronto
Toronto, Ont. M5G 1X5 (Canada)
Tel. +1 416 586 4800, ext. 2859, Fax +1 416 586 8588, E-Mail quaggin@mshri.on.ca

express VEGF-A when they are fully differentiated, although absolute levels of expression decrease. The adjacent and intimately associated glomerular endothelial cells express the cognate tyrosine kinase receptors for VEGF-A – named VEGFR-1 (flt1) and VEGFR-2 (flk-1). As the glomerulus forms, a vascular cleft develops between the podocyte precursors and presumptive proximal tubule; VEGFR-2-positive glomerular endothelial cells migrate into this vascular cleft to form the glomerular capillaries [5, 6].

A number of animal studies have been performed that demonstrate VEGF-A signaling is key for proper formation of the GFB. Newborn rats or mice injected with VEGF-A aptamers that inhibit VEGF-A signaling exhibit smaller glomeruli with fewer capillary loops [7]. VEGF-A knockout mice die at 9.5 days of embryonic life (days post-coitum; dpc) due to major disruption of vascular development; mice that are haploinsufficient for the VEGF-A gene also die during embryonic life at 10.5 dpc due to major vascular defects. These results demonstrate dosage sensitivity for VEGF-A in angiogenesis [1]. The early embryonic lethality precludes the analysis of the role of VEGF-A signaling in the kidney, as glomerular development does not begin until 13.5 dpc. To overcome this hurdle, a conditional Cre-loxP targeting strategy was used [9].

Experiments were performed to determine if there is dosage sensitivity to VEGF-A in the developing glomerulus, similar to the dosage sensitivity observed for vascular development in the whole embryo. Podocyte-selective deletion of both VEGF-A genes from transgenic mice results in perinatal lethality and renal failure [9]. The glomeruli of these mice fail to form functional filtration barriers due to major defects in endothelial cell migration, survival and differentiation. In mature glomeruli, no endothelial cells are found. In turn, mesangial cells fail to migrate into the developing glomeruli most likely due to the absence of endothelial-derived growth factors such as PDGF-B [10, 11].

Loss of a single VEGF-A allele in podocytes leads to endotheliosis – the renal lesion observed in pre-eclampsia and other microangiopathies. The mice exhibit proteinuria by 2.5 weeks of age that progresses to nephrotic range proteinuria accompanied by renal failure, glomerulosclerosis and death at 9–12 weeks of age [9]. Mice that carry a single hypomorphic VEGF-A allele in their podocytes (i.e. one allele is deleted from the podocytes, and the remaining allele is hypofunctional) die at 3 weeks of age from renal failure [12]. Although glomeruli develop normally, glomerular endothelial cells disappear within 1–3 days following glomerular maturation, leaving behind capillary ghosts. A similar phenomenon has been observed in tumors following treatment with anti-VEGF therapy, where the tumor vessels are lost but basement membrane sheaths remain [13]. Following the disappearance of endothelial cells, the mesangial compartment undergoes rapid mesangiolysis. By contrast, overexpression of the major 164 isoform of VEGF-A to mRNA levels 15- to 20-fold higher than in wild-type mice, leads to collapse of the glomerular tuft, proteinuria and death from renal failure within the first week of life [9].

Taken together, these results provide the following insights: local VEGF-A signaling is required for glomerular development; exquisite dosage sensitivity to the VEGF-A gene exists during glomerulogenesis; finally, and perhaps most surprisingly, each dose reduction of the VEGF-A gene is associated with a distinct, and dramatic glomerular pathology and clinical course. However, despite the differences, all of the injuries begin within the glomerular endothelial compartment; more severe reductions in VEGF-A dose result in more severe endothelial defects and lead to the diverse pathologies (fig. 1; table 1).

Are All VEGF-A Isoforms Created Equally?

Multiple splice variants of the VEGF-A gene give rise to a number of pro- and anti-angiogenic isoforms. As these isoforms have different properties, they might possess combined as well as unique functions. The VEGF-A gene contains 8 exons; the heparin-binding domain of VEGF-A is encoded by the sixth and seventh exons. The three major isoforms (VEGF188/9, VEGF164/5 and VEGF120/1) are generated by alternative splicing from exon 5 to exon 6, from exon 5 to exon 7, or from exon 5 to exon 8, respectively. The 120/1 isoform lacks the heparin-binding domain and is freely secreted, whereas a large part of 164/5 and most of 188/9 remain in the extracellular matrix surrounding the source. For each major isoform, a 'b' isoform exists that arises from an alternative distal splice site in exon 8 [14]. This results in isoforms of the same size but with an alternate carboxyl-terminus. Investigators have shown that the 165b isoform can inhibit VEGF165-mediated endothelial cell proliferation (HUVEC cells) in vitro [15].

The relative contribution of the major and minor isoforms of VEGF-A in development and disease are not yet entirely clear. However, valuable insight has been gained through targeted modification of the VEGF-A locus, forcing the VEGF-A gene to express only one of the major

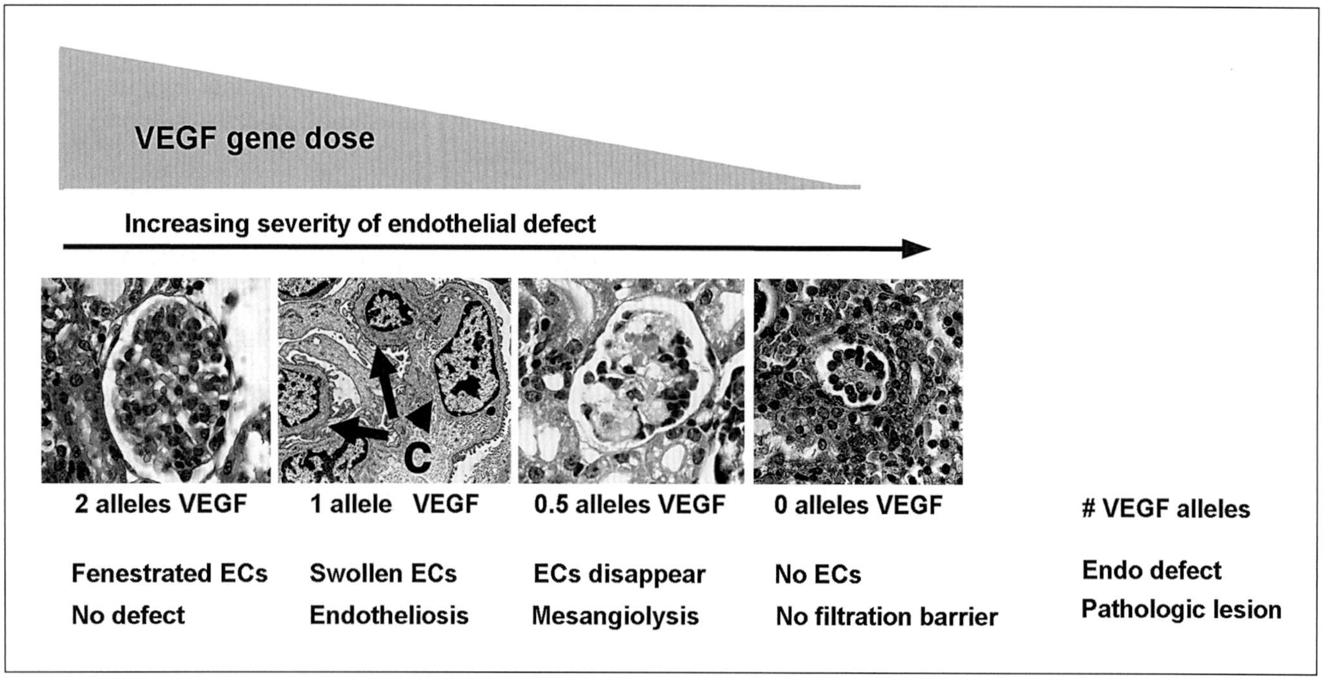

Fig. 1. Single gene defect. Multiple renal lesions.

Table 1. VEGF-A dose determines severity of endothelial injury

Podocyte dose of VEGF-A (number of VEGF-A alleles)	Endothelial defects	Pathology and clinical course
No VEGF-A in podocyte (0 alleles; 0%)	No glomerular endothelium in glomeruli; no GFB	No GFB forms; no urine
Hypomorphic model (1 hypofunctional VEGF allele in podocytes) (0.5 alleles; approx. 25%)	Endothelial cells migrate into glomeruli but disappear by 1–3 days postnatal	Massive mesangiolysis and ESRD by 3 weeks of age
1 VEGF-A allele in podocytes (1 allele; 50%)	GFB development is OK; endothelial cells swell by 2.5 weeks (endotheliosis)	Glomerulosclerosis and ESRD by 9–12 weeks of age
Wild type (2 alleles; 100%)	Fenestrated endothelium; normal GFB	Healthy
Increased VEGF-164 in podocytes (15–20-fold) (multiple alleles; >100%)	Endothelial cells are present but capillary loops are collapsed; hypermeability of GFB	Collapsing glomerulopathy with nephrotic proteinuria and ESRD by 5 days of age

isoforms in transgenic mice. Mice that express only the 120 isoform exhibit multiple defects in bone, neural, cardiovascular and renal development. In the kidney, there is a 25% reduction in the total number of glomeruli but a 40% reduction in the number of vascularized glomeruli [16]. Mice that express only the 188 isoform show, among other things, impaired arterial development in the retina, but have no reported glomerular defects. To date, no phenotype has been reported for the '164-isoform-only' mouse – suggesting that it really might be the 'master' isoform and is sufficient for VEGF-A function, although this remains speculation at present.

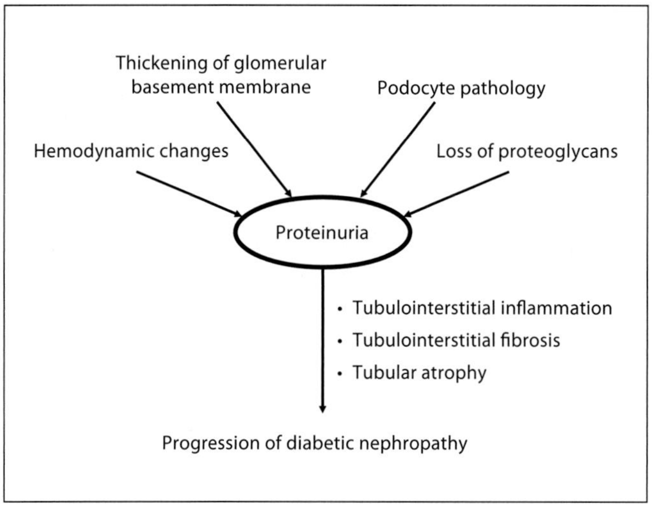

Fig. 2. Overview of mechanisms leading to proteinuria in diabetic nephropathy. Although hemodynamic factors (e.g. glomerular hypertension and hyperfiltration), thickening of the GBM, and loss of negatively charged proteoglycans are important factors for the development of proteinuria, recent research has focused on the primary role of podocyte pathology in this process (for details see text). Proteinuria induces in tubular cells a cascade of events that leads to interstitial inflammation/fibrosis and tubular atrophy.

tory proteins p57 and $p27^{Kip1}$, apoptotic loss of cells cannot be replaced [20]. High glucose additionally stimulates the expression of $p27^{Kip1}$ resulting in cell cycle arrest before successful replication (fig. 1). We found recently that AGEs also stimulate the expression of $p27^{Kip1}$ in podocytes, likely enhancing the maladaptive phenotype leading to apoptosis [Wolf et al., unpubl. observations]. $p27^{Kip1}$ knockout mice are partially protected from the development of diabetic nephropathy, and morphometric analysis indicates that there is less podocyte injury in diabetic $p27^{Kip1}$ knockout mice compared with diabetic wild-type animals, suggesting an important role of $p27^{Kip1}$ in podocyte injury [20].

Apoptosis of podocytes can further enhance proteinuria. Susztak et al. [21] discovered that glucose-mediated ROS generation, which is a common upstream initiator of many pathophysiological abnormalities in diabetes mellitus, is responsible for podocyte depletion early on in experimental diabetic nephropathy. ANG II contributes to renal ROS formation through activation of NADPH oxidase. Moreover, depletion of podocytes allows the denuded GBM to come into close contact with Bowman's capsule, resulting in synechiae formation, the first step in the development of glomerulosclerosis.

Conclusion

Proteinuria, an important contributor to progression of diabetic nephropathy, is caused by several factors. On a molecular level, metabolic pathways activated by hyperglycemia as well as proteins altered by high blood glucose such as Amadori-albumin and AGEs are key players in the development of diabetic nephropathy (fig. 2). Oxidative stress is an early event in the development of diabetic complications. A variety of growth factors and cytokines are then induced through complex signal transduction pathways. The cytokine TGF-β1 has emerged as an important downstream mediator for the development of renal hypertrophy and the accumulation of extracellular matrix components. Besides changes in the GBM with thickening and loss of negatively charged sites that contribute to increase flux of proteins through the glomerular ultrafiltration barrier, recent research has focused on podocyte pathology as a major factor causing proteinuria in diabetic nephropathy. Podocytes are a victim of pathophysiological events associated with diabetic nephropathy and are also targets of the effects of high glucose, AGEs, mechanical stretch, and proteinuria itself. A reduction in nephrin protein at the slit diaphragm is associated with podocyte foot process effacement. Proteinuria itself can then induce tubular atrophy and interstitial fibrosis. Podocyte injury is thus a culprit for progression of diabetic nephropathy. Many of these events are mediated by ANG II whose local concentration could be stimulated by high glucose, mechanical stretch, and proteinuria. ANG II, in turn, stimulates VEGF and suppresses nephrin expression events that can aggravate the proteinuria, and also induces TGF-β1 expression leading ultimately to an increase in podocyte apoptosis and fostering the development of glomerulosclerosis.

Acknowledgements

G.W. is supported by a grant from the Deutsche Forschungsgemeinschaft. F.N.Z. is supported by grants from the American Diabetes Association and the National Institutes of Health (DK44513 and DK54608).

strated that chymase, an ANG II-forming enzyme not inhibited by ACE inhibitors, is upregulated in glomeruli of patients with nephropathy due to type 2 diabetes. This suggests that local glomerular formation of ANG II in the diabetic state may be partly independent of ACE and is not abolished by ACE inhibition alone. As a consequence, an increase in local ANG II in the podocyte microenvironment suppresses nephrin expression, and thereby enhances ultrafiltration of proteins that further contributes to podocyte damage [13]. Interestingly, recent data suggest that treatment with the glitazones prevents the downregulation of nephrin in the diabetic state [14].

A loss of nephrin and proteoglycan expression may not be the only pathological change fostering proteinuria in diabetic nephropathy. VEGF is an angiogenic factor that markedly increases vascular permeability [12]. The expression of VEGF is pronounced in podocytes, and experimental diabetes increases VEGF mRNA and protein expression. Exposure of cultured podocytes to high glucose concentrations also upregulates VEGF expression [15]. Also, ANG II appears to be an important mediator of the diabetes-associated increase in VEGF synthesis in podocytes because an AT1-receptor antagonist attenuates these changes [15]. In addition, ANG II directly stimulates VEGF expression in cultured podocytes even under normal glucose conditions [15]. ANG II-mediated VEGF production in the kidney is not only mediated by AT1- but also AT2-receptors [12]. In fact, only an ACE inhibitor, but not an AT1-receptor antagonist, attenuated enhanced renal VEGF protein concentrations in streptozotocin-induced diabetic rats [12]. An important stimulatory transcription factor for VEGF expression is hypoxia-inducible factor-1 (HIF-1). We have recently shown that ANG II stimulates HIF-1α expression by a posttranscriptional mechanism via AT2-receptors [12]. This increase is caused by downregulation of SM-20/PHD3, the enzyme responsible for hydroxylation of HIF-1α initiating degradation [12].

Rincon-Choles et al. [16] recently discovered that expression of ZO-1, a junctional protein involved in the assembly of tight junctions and expressed at the podocyte slit diaphragm, is reduced in animal models of types 1 and 2 diabetes. Exposure of podocytes to high glucose in vitro also resulted in a decrease in ZO-1 expression [16]. This effect can contribute to alterations in podocyte structure and function that enhance proteinuria.

Proteinuria itself could amplify podocyte injury further by a positive feedback cycle. In rats with reduced renal mass, plasma protein accumulation in podocytes,

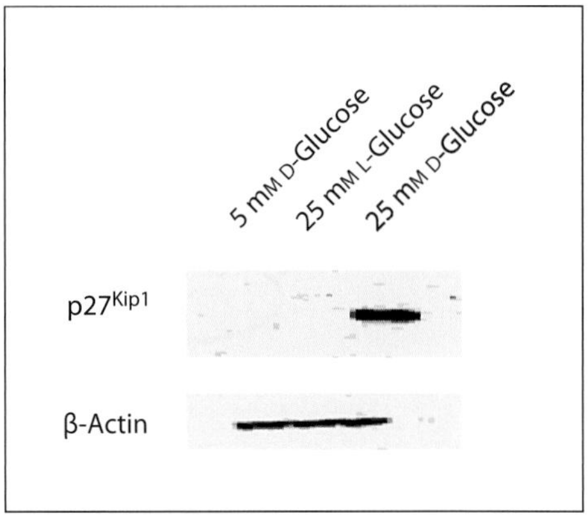

Fig. 1. Induction of $p27^{Kip1}$ by high glucose in cultured podocytes. This Western blot shows induction of the cell cycle inhibitor $p27^{Kip1}$ by 25 mM D-glucose compared with normal glucose (5 mM D-glucose). This effect is independent of osmolarity because 25 mM L-glucose, which is not taken up by cells, has no effect. Upregulation of $p27^{Kip1}$ in podocytes sensitizes these cells for hypertrophy and subsequent apoptosis.

which is a consequence of enhanced transcapillary passage, induces the profibrotic cytokine TGF-β1 [17]. Both albumin and high ambient glucose stimulate TGF-β1 expression in cultured podocytes [12]. An increase in TGF-β1 can stimulate extracellular matrix synthesis by the podocyte itself in an autocrine manner or may reach mesangial cells to induce a sclerosing phenotype in these cells. Advanced glycation end-products (AGEs) are also involved in induction of TGF-β1 in podocytes. Expression of RAGE, the receptor for AGEs, is increased in podocytes in diabetic *db/db* mice [12]. Activation of RAGE leads to increased VEGF and TGF-β1 expression [12]. We recently reported that blockade of VEGF signaling by a receptor kinase inhibitor ameliorates diabetic albuminuria in *db/db* mice [18]. These in vivo data support a link between long-term metabolic events in diabetes such as AGEs to increased expression of VEGF and TGF-β1 and eventually to podocyte pathology that manifests in proteinuria.

An increase in TGF-β1 may have additional effects besides stimulation of extracellular matrix. In podocytes, high levels of TGF-β1 induce apoptosis through Smad7 by inhibiting nuclear translocation of the cell survival factor NF-κB [19]. TGF-β1-mediated activation of Smad7 appears to be specific for podocytes and is not found in mesangial cells, as assessed in a limited biopsy series of diabetic nephropathy patients [19]. Since podocytes are terminally differentiated cells that are unable to undergo cell division because of upregulation of cell cycle inhibi-

The glomerular basement membrane (GMB) contains a number of heparan-binding proteins and likely functions as a 'sink' for the storage and perhaps release of VEGF-A isoforms that can bind heparan sulfates. Damage to the GBM might in turn lead to diminished VEGF-A stores with reduced ability to repair and maintain a healthy glomerular endothelial layer.

Despite the myriad of papers that have documented altered levels of VEGF-A in disease or animal models, relatively few have dissected the contribution of different isoforms [17]. Given the existence of inhibitory isoforms, this clearly becomes an important issue – as global changes in mRNA or protein levels may not reflect VEGF-A signaling function at all. Further dissection of the individual and combined roles of VEGF-A isoforms remains a challenging but worthy task.

What Regulates Glomerular VEGF-A Expression?

Improved understanding of how VEGF-A is regulated in the glomerulus is also a priority. One of the most perplexing questions regarding VEGF-A biology is the observed dosage sensitivity. VEGF-A is an inducible gene; inductions of 3-fold or greater have been reported following hypoxic stimulation [8]. Why doesn't a wild-type VEGF-A allele compensate for loss of a single allele? Some investigators argue that induction does occur in the haploinsufficient models, but it just does not happen fast enough. This remains unproven.

Although hypoxia and increased transcription from the VEGF-A gene through stabilization of HIF alpha subunits is a well-known stimulus for VEGF-A induction, the healthy mature glomerulus is not hypoxic [18]. Factors responsible for maintenance of constitutive expression of VEGF-A in the filtering glomerulus are intriguing. In vitro studies have suggested nonhypoxic regulation through integrins [19], although the corresponding $\alpha 3$ integrin knockout mice do not exhibit decreased VEGF expression [Kreidberg, pers. commun.]. The Wilms' tumor suppressor gene, a zinc finger-containing transcription factor that is expressed in developing and mature podocytes, can also upregulate VEGF expression in metanephric mesenchymal cells (precursors of podocytes and tubular nephron) in organ culture [20], but whether this affects expression in the podocyte is unknown.

Podocytes exposed to high glucose or angiotensin II also upregulate VEGF-A production in vitro [21, 22]. These same stimuli are known to lead to increased levels of HIFs that in turn regulate production of VEGF-A. The link between glucose and VEGF-A and reports of elevated VEGF-A levels in diabetic nephropathy (DN) have suggested a causal role for VEGF-A in the pathogenesis of this renal disease [23–25]. However, it is possible that the elevated VEGF-A levels simply represent a 'rebound' phenomenon to damaged or dysregulated endothelium (i.e. the glomeruli and podocytes are hypoxic in DN leading to induction of VEGF-A as a secondary phenomenon). Sorting out what factors regulate VEGF-A in the glomerulus will provide important clues to the role of VEGF-A in human pathology.

Paracrine, Autocrine and Endocrine Functions of VEGF-A Signaling in the Glomerulus

Results from conditional gene targeting experiments demonstrate a major paracrine signaling pathway for VEGF-A within the glomerulus as the most profound and earliest defects are observed in the endothelium in all cases. However, there is recent in vitro evidence to suggest that VEGF-A may also function in an autocrine fashion that is required for podocyte survival [26, 27].

Podocytes express a number of VEGF-A receptors including VEGFR1, VEGFR3, neuropilin-1 and neuropilin-2 in vitro. Conflicting reports have been published regarding the expression of VEGFR-2 in podocytes [26–28]. This is a key issue, as the majority of VEGF-A signaling is believed to occur through this receptor. Elegant studies by Foster and colleagues have convincingly shown that podocytes in culture respond to a VEGF-mediated signal; podocyte survival is diminished upon inhibition of VEGF receptors [26, 29]. The tyrosine kinase inhibitor used in the studies inhibits VEGFR2, VEGFR1 and VEGFR3. The same investigators have shown that podocytes in culture also respond to VEGF-C [30], which is expressed in the glomerulus in vivo [31]. Definitive studies to test this possibility might include deletion of the VEGF receptors from podocytes in vivo. We have generated transgenic mice with genetic deletion of the VEGFR-2 receptor from podocytes. Some of the mice develop late-onset focal and segmental glomerulosclerosis, but the phenotype is variable. This might reflect incomplete DNA excision or unexplained stochastic events and requires additional analysis before conclusions can be drawn.

Role of VEGF-A in Maintenance of Glomerular Function

The strongest evidence that VEGF-A plays a role in maintenance of the glomerular endothelium comes from a series of beautiful studies from Ananth Karumanchi's group in Boston; they clearly demonstrated an association between the circulating levels of sFlt1, a truncated version of the VEGFR-1 protein that is missing the cytoplasmic kinase tail, and pre-eclampsia, a clinical syndrome of pregnancy characterized by proteinuria, endotheliosis and hypertension [32]. Elevated levels of sFlt1 correlate inversely with the circulating levels of free VEGF-A and directly with the severity of pre-eclampsia [33]. Furthermore, rats or mice injected with a synthetic sFlt molecule, or expressing a sFlt lentivirus, also develop endotheliosis, hypertension and proteinuria [32, 34]. Genetic deletion of a single VEGF allele from podocytes results in endotheliosis and proteinuria in mice as discussed above [9].

In the glomerular lesion of pre-eclampsia, the fenestrations of glomerular endothelial cells are lost, which is associated with proteinuria. How do the elevated levels of sFlt lead to the glomerular lesion? Titration of circulating free VEGF-A may not be the entire story. Perhaps this soluble receptor is able to cross the filtration barrier and/or is capable of titrating locally produced VEGF-A; this is still an area of discussion. In interpreting the glomerular lesions, it is important to recognize that some of the sFlt- or VEGF-blocking compounds used in animal studies contain the active VEGF-binding site, but do not contain all of the Ig repeats present in the endogenous sFlt gene product. It is possible that some of these pharmacologic or synthetic compounds pass more readily across the GFB than endogenous sFlt. Complementary genetic studies to delete VEGF-A from mature glomeruli of transgenic mice are ongoing; preliminary results support a model where function of the GFB in adults depends upon VEGF-A production in the podocyte [Quaggin, unpubl.].

In addition to the renal lesion of pre-eclampsia, many studies report altered glomerular levels of VEGF-A protein and/or mRNA in renal disease. These range from diabetic nephropathy to the microangiopathies associated with various medications and primary glomerulopathies. In these studies, loss of VEGF-A expression may represent true down regulation of the gene and its product or alternatively, loss or dysregulation of podocytes that produce VEGF-A [35]. The challenge in interpretation stems from the complexity of the VEGF gene, inadequacies in detecting the mRNA and protein and the scarcity of renal tissue at specific stages of a disease. It should be emphasized that although imperfect, expression studies in patient samples are imperative first steps to dissect the role of VEGF-A in human disease.

Conclusions

The evidence is clear that VEGF signaling is essential for the formation and maintenance of a functional glomerular filtration barrier. This has direct implications for patients in the clinic who are receiving anti-VEGF therapies; clinicians should be aware of the role of VEGF-signaling in the kidney. Baseline tests for proteinuria and renal function should be performed prior to starting anti-angiogenic therapy and continued over the course of therapy. The challenge for researchers in the next few years will be to dissect the mechanisms of regulation of VEGF, determine functions of various combinations of VEGF-A isoforms, and understand the relevance of various routes of VEGF-A signaling, i.e. paracrine versus autocrine versus endocrine. These are important factors to consider and studies to pursue if VEGF-A is to be used as a prognostic marker and/or therapeutic tool in our patients.

References

1 Carmeliet P, et al: Abnormal blood vessel development and lethality in embryos lacking a single VEGF allele. Nature 1996;380:435–439.
2 Benjamin LE, Hemo I, Keshet E: A plasticity window for blood vessel remodelling is defined by pericyte coverage of the preformed endothelial network and is regulated by PDGF-B and VEGF. Development 1998;125:1591–1598.
3 Ferrara N, Gerber HP: The role of vascular endothelial growth factor in angiogenesis. Acta Haematol 2001;106:148–156.
4 Gerber HP, et al: VEGF is required for growth and survival in neonatal mice. Development 1999;126:1149–1159.
5 Abrahamson DR: Glomerulogenesis in the developing kidney. Semin Nephrol 1991;11:375–389.
6 Hyink DP, Abrahamson DR: Origin of the glomerular vasculature in the developing kidney. Semin Nephrol 1995;15:300–314.
7 Ostendorf T, et al: VEGF(165) mediates glomerular endothelial repair. J Clin Invest 1999;104:913–923.
8 Pierce EA, et al: Vascular endothelial growth factor/vascular permeability factor expression in a mouse model of retinal neovascularization. Proc Natl Acad Sci USA 1995;92:905–909.

9 Eremina V, et al: Glomerular-specific alterations of VEGF-A expression lead to distinct congenital and acquired renal diseases. J Clin Invest 2003;111:707–716.
10 Leveen P, et al: Mice deficient for PDGF B show renal, cardiovascular, and hematological abnormalities. Genes Dev 1994;8:1875–1887.
11 Lindahl P, et al: Paracrine PDGF-B/PDGF-Rbeta signaling controls mesangial cell development in kidney glomeruli. Development 1998;125:3313–3322.
12 Eremina V, et al: VEGF-A signaling in the podocyte-endothelial compartment is required for mesangial cell migration and survival. JASN 2006;17:724–735.
13 Mancuso MR, et al: Rapid vascular regrowth in tumors after reversal of VEGF inhibition. J Clin Invest 2006;116:2610–2621.
14 Bates DO, et al: VEGF165b, an inhibitory splice variant of vascular endothelial growth factor, is down-regulated in renal cell carcinoma. Cancer Res 2002;62:4123–4131.
15 Bates DO, Harper SJ: Therapeutic potential of inhibitory VEGF splice variants. Future Oncol 2005;1:467–473.
16 Mattot V, et al: Loss of the VEGF(164) and VEGF(188) isoforms impairs postnatal glomerular angiogenesis and renal arteriogenesis in mice. J Am Soc Nephrol 2002;13:1548–1560.
17 Bortoloso E, et al: Quantitative and qualitative changes in vascular endothelial growth factor gene expression in glomeruli of patients with type 2 diabetes. Eur J Endocrinol 2004;150:799–807.
18 Freeburg PB, et al: Podocyte expression of hypoxia-inducible factor (HIF)-1 and HIF-2 during glomerular development. J Am Soc Nephrol 2003;14:927–938.
19 Datta K, et al: Regulation of vascular permeability factor/vascular endothelial growth factor (VPF/VEGF-A) expression in podocytes. Kidney Int 2004;66:1471–1478.
20 Gao X, et al: Angioblast-mesenchyme induction of early kidney development is mediated by Wt1 and Vegfa. Development 2005;132:5437–5449.
21 Kang YS, et al: Angiotensin II stimulates the synthesis of vascular endothelial growth factor through the p38 mitogen activated protein kinase pathway in cultured mouse podocytes. J Mol Endocrinol 2006;36:377–388.
22 Iglesias-de la Cruz MC, et al: Effects of high glucose and TGF-beta1 on the expression of collagen IV and vascular endothelial growth factor in mouse podocytes. Kidney Int 2002;62:901–913.
23 Hovind P, et al: Elevated vascular endothelial growth factor in type 1 diabetic patients with diabetic nephropathy. Kidney Int 2000;75(suppl):S56–S61.
24 Chiarelli F, et al: Vascular endothelial growth factor (VEGF) in children, adolescents and young adults with type 1 diabetes mellitus: relation to glycaemic control and microvascular complications. Diabet Med 2000;17:650–656.
25 McLaren M, et al: Elevated plasma vascular endothelial cell growth factor and thrombomodulin in juvenile diabetic patients. Clin Appl Thromb Hemost 1999;5:21–24.
26 Foster RR, et al: Functional evidence that vascular endothelial growth factor may act as an autocrine factor on human podocytes. Am J Physiol 2003;284:F1263–F1273.
27 Guan F, et al: Autocrine VEGF-A system in podocytes regulates podocin and its interaction with CD2AP. Am J Physiol 2006;291:F422–F428.
28 Chen S, et al: Podocyte-derived vascular endothelial growth factor mediates the stimulation of alpha3(IV) collagen production by transforming growth factor-beta1 in mouse podocytes. Diabetes 2004;53:2939–2949.
29 Foster RR, et al: Vascular endothelial growth factor and nephrin interact and reduce apoptosis in human podocytes. Am J Physiol 2005;288:F48–F57.
30 Foster RR, et al: VEGF-C promotes survival in podocytes. Am J Physiol 2006;291:F196–F207.
31 Partanen TA, et al: VEGF-C and VEGF-D expression in neuroendocrine cells and their receptor, VEGFR-3, in fenestrated blood vessels in human tissues. FASEB J 2000;14:2087–2096.
32 Maynard SE, et al: Excess placental soluble fms-like tyrosine kinase 1 (sFlt1) may contribute to endothelial dysfunction, hypertension, and proteinuria in preeclampsia. J Clin Invest 2003;111:649–658.
33 Levine RJ, et al: Circulating angiogenic factors and the risk of preeclampsia. N Engl J Med 2004;350:672–683.
34 Sugimoto H, et al: Neutralization of circulating vascular endothelial growth factor (VEGF) by anti-VEGF antibodies and soluble VEGF receptor 1 (sFlt-1) induces proteinuria. J Biol Chem 2003;278:12605–12608.
35 Baelde HJ, et al: Reduction of VEGF-A and CTGF expression in diabetic nephropathy is associated with podocyte loss. Kidney Int 2007.

nephron
Clinical Practice

Nephron Clin Pract 2007;106:c61–c66
DOI: 10.1159/000101799

Published online: June 6, 2007

Podocyte Damage Resulting in Podocyturia: A Potential Diagnostic Marker to Assess Glomerular Disease Activity

Arndt Petermann[a, b] Jürgen Floege[a]

[a]Division of Nephrology, RWTH University of Aachen, Aachen, and [b]Nephrology Centers of Villingen-Schwenningen, Villingen-Schwenningen, Germany

Key Words
'Crossing the bridge' · Podocyte detachment · Glomerulonephritis · Proteinuria · Apoptosis

Abstract
A decrease in podocyte number contributes to the development of glomerulosclerosis in most forms of glomerular disease [1–5]. Traditionally, it has been argued that this decrease may be caused by the inability of podocytes to proliferate and replace those lost following immune, metabolic, toxic or hemodynamic injury. These data contrast with recent studies showing that podocytes are able to enter the cell cycle after injury, to progress through the different phases of the cell cycle and even enter mitosis. However, experimental and human data suggest that entry of podocytes into the cell cycle may result in reduced adhesion to the glomerular basement membrane with subsequent loss of podocytes into the urine and excretion of both viable and apoptotic podocytes. Viable urinary podocytes can be cultivated ex vivo for up to 2–3 weeks and in experimental models precede the onset of proteinuria. More importantly, podocyturia can decrease despite persistent proteinuria. The latter observation suggests that podocyturia may serve as the first non-invasive marker of 'active' glomerular damage and might thus drive therapeutic interventions in the future. However, at present technical issues still prevent a broad clinical application of podocyturia detection in clinical practice.

Podocyte damage is central in the development of glomerulosclerosis. Both progressive experimental and human renal disease is accompanied by podocytopenia [1–6]. One of the earliest signs of podocyte damage is often the loss of the normal differentiated cell morphology with effacement of the foot processes. Foot process effacement may manifest as partial retraction up to total loss of foot processes. Electron-microscopic studies have suggested that podocyte damage resulting in foot process effacement or fusion or even cell death may lead to focal denudation of the glomerular basement membrane (GBM). Denuded GBM caused by podocyte damage leads to its outward bulging, given the high intraglomerular pressure, and such denudations are then the origin of focal adhesions of the glomerular tuft to the outer leaflet of Bowman's capsule [7]. Spreading of the lesion to a focal scar follows with misdirected filtration into the renal interstitium, and ultimately spreading of primary glomerular injury into the renal tubulointerstitium [7]. While the above may appear as a rather uniform response pattern of podocytes to injury, podocyte damage can actually have a number of further consequences (table 1). Key mechanisms include hypertrophy, cell proliferation, detachment, and cell death.

KARGER

Fax +41 61 306 12 34
E-Mail karger@karger.ch
www.karger.com

1660–2110/07/1062–0061$23.50/0

Accessible online at:
www.karger.com/nec

Jürgen Floege, MD
Division of Nephrology and Clinical Immunology, RWTH Aachen
Pauwelstrasse 30
DE–52057 Aachen (Germany)
Tel. +49 241 8089 530, Fax +49 241 8082 446, E-Mail juergen.floege@rwth-aachen.de

Table 1. Podocyte responses to injury

Morphological outcome after podocyte injury	Mechanism
Restitution of normal glomerular architecture	Transient change in podocyte morphology and reconstitution of normal phenotype Replacement of damaged podocytes by extra- or intrarenal stem cells or by other intrarenal cells (e.g. parietal epithelial cells)?
Glomerular hypertrophy	Podocyte cell division with maintenance of differentiated phenotype (probably does not occur in adult life) Podocyte hypertrophy including potential nuclear division with formation of bi- and multinucleated podocytes
Focal segmental glomerulo-sclerosis (FSGS)	Podocyte loss from the GBM with subsequent excretion of viable podocytes in the urine, podocyte apoptosis or necrosis Podocyte apoptosis or necrosis in situ Podocyte dedifferentiation with subsequent proliferation and/or migration or loss from the GBM (as observed in collapsing FSGS or crescentic glomerulonephritis)

Podocyte Responses to Injury: Hypertrophy, Proliferation, Detachment and Death

During nephrogenesis podocytes develop from a highly proliferative cuboidal cell layer in the S-shaped bodies. At the transition to the capillary loop stage podocytes lose their capacity to divide and differentiate to their adult phenotype, producing interdigitating processes. This may relate to a switch in cell cycle protein expression, e.g. from cyclin D1 to cyclin D3 together with expression of CDK inhibitors [8].

Perhaps the most important mechanism whereby podocytes respond to damage, increased intraglomerular pressure and denuded GBM areas is via cellular hypertrophy [9]. Indeed, experiments in cultured podocytes demonstrated that a switch from cell proliferation to hypertrophy induced by mechanical stretch is mediated by overexpression of the cell-cycle inhibitor p21 [10].

Following glomerular damage, e.g. in passive Heymann nephritis (PHN), a model of membranous nephropathy, podocytes can re-enter the cell cycle including the development of mitotic figures resembling prophase, metaphase or anaphase [11, 12]. However, whereas nuclear division, albeit rare, is well documented in podocytes and bi- or multinucleated podocytes have been detected, it is much less certain whether full cell division can occur in situ. Theoretically, the key dilemma of a podocyte attempting cell division is that normally cells round up to do so and reduce their strength of attachment to the underlying surface. In the case of the podocyte this would have to occur in the presence of high or often increased intraglomerular pressure, considerable stretch forces and constant filtrate flow. Thus, attempted cell division may be a key mechanism that leads to the detachment of podocytes from the GBM.

Data from TGF-β transgenic mice support a role of apoptosis in the development of glomerulosclerosis [13]. Stretch of cultured podocytes also induced cell death, which was prevented by blockage of angiotensin II [14]. Further, podocyte apoptosis may result from disturbances of the slit membrane, since CD2AP knockout mice showed a significant increase in podocyte apoptosis as compared to wild-type mice [15] and since intact nephrin also contributes to podocyte survival [16]. Not surprisingly, the underlying matrix determines survival as well, and podocytes grown on collagen exhibit an increased survival in comparison to those grown on plastic substratum [17].

Thus, once podocytes detach from the GBM following either injury with loss of adhesion or reduced adhesion during the attempt to undergo cell division they will be lost into the urine. Such detached podocytes likely have a high risk for apoptotic or necrotic cell death and this may be one of the origins of podocyte-specific proteins or mRNA found in urine of patients with glomerulonephritis [18]. In addition to cell death, there are now numerous reports demonstrating that podocytes with preserved surface structures or even viable podocytes can also be detected in the urine.

Podocytes in Urine

Cells expressing podocalyxin, and therefore assumed to be podocytes, have repeatedly been described in cytospins of morning urine of patients with FSGS, diabetic nephropathy or IgA nephropathy but not in patients with minimal change or membranous nephropathy [19–21]. However, the viability of these cells was not determined. The extent of podocyturia was reduced by therapeutic interventions such as pioglitazone or antiplatelet drugs in type 2 diabetic patients [22, 23], cerivastatin or renin-angiotensin system blockers in patients with mesangioproliferative glomerulonephritis [24, 25]. In addition to intact podocytes, apical cell membranes expressing podocalyxin were also detected in patient urine [26].

In contrast to the above data, a very recent study, using an antibody to synaptopodin and urine samples of 100 highly proteinuric patients, detected podocyturia in only 1 patient [27]. This latter study therefore questioned whether relying on podocalyxin as a single parameter is specific enough for podocytes in the urine. Indeed podocalyxin can also be detected on platelets and vascular endothelia.

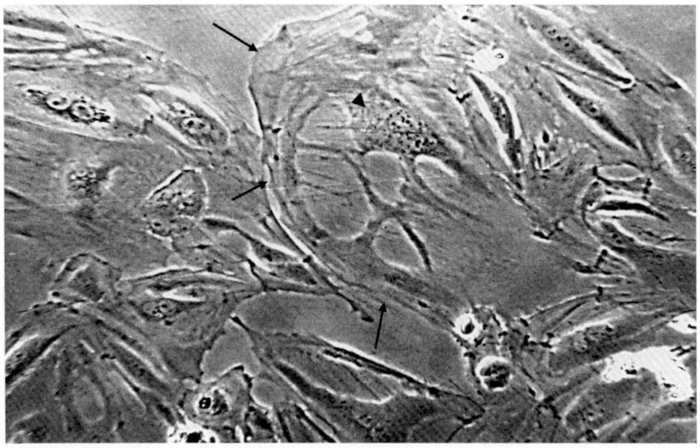

Fig. 1. Podocytes cultured from the urine of rats 5 days after induction of passive Heymann nephritis, a model of membranous glomerulopathy. At 10 days of ex vivo culture, the cells exhibit a typical flat morphology, with a belt-like outer cytoplasm (arrow) that is connected with the central body (arrowhead) via small cytoplasmic extensions.

Cultivation of Viable Podocytes from Urine

To determine whether viable podocytes are lost from damaged glomeruli, we examined urine of rats with PHN. Urine samples were centrifuged, resuspended in tissue culture media, seeded onto collagen-coated tissue culture plates and then grown under typical cell culture conditions. Adherent cells showed a characteristic morphology and often were bi- or multinucleated (fig. 1). They frequently exhibited mitotic figures and they were positive for synaptopodin, nephrin, podocin, WT-1 and GLEPP1 (podocyte specific antigens) and negative for mesangial (Thy 1.1), tubular (Tamm-Horsfall protein) and endothelial (RECA) markers [28]. Podocyte numbers ex vivo increased under these conditions during the first 5 days and cell cultures uniformly stained positive for podocyte specific proteins. Thereafter, apoptosis increased and cell numbers decreased again. Podocyte detachment was not uniform during PHN but rather reached a maximum around day 5, i.e. 2 days after the onset of proteinuria. Very similar findings were obtained within 24 h of inducing type I diabetes in rats but not in control rats [28]. Again, the cell number ex vivo increased daily for about 4 days, which is consistent with cell proliferation.

Podocyturia as a Marker for Disease Activity

In the studies of Hara et al. [19] podocyte excretion in patients with glomerulonephritis appeared to correlate with disease activity as determined by the extent of proteinuria and hematuria. Taken together with the effects of therapeutic interventions on podocyturia (see above), this suggested that it might reflect glomerular disease activity. To better assess this issue, we studied the urinary excretion of viable podocytes in two models of transient glomerular injury, i.e. rats with puromycin aminonucleoside-induced nephrosis (PAN) and mesangioproliferative anti-Thy 1.1 glomerulonephritis, as well as in a model of continuous glomerular injury, i.e. hypertensive nephropathy following 5/6-nephrectomy, and finally in ageing rats [29].

Urinary podocyte loss occurred in PAN prior to the onset of proteinuria. More importantly, proteinuria subsequently persisted despite remission of podocyturia. Podocyturia also occurred in anti-Thy 1.1 nephritis, where it again antedated proteinuria and fell rapidly despite ongoing proteinuria. In this model, the primary injury is highly selective for mesangial cells leading to their destruction followed by capillary dilatation and intraglomerular microaneurysms after 3–4 days. All of this will contribute to secondary podocyte damage via altered physical forces, e.g. stretching [7], and/or potential biochemical alterations of the normal GBM in the course of the disease. In continuous glomerular injury, i.e. following 5/6 nephrectomy, podocyturia never subsided and rather mirrored proteinuria as well as systemic hypertension, whereas no podocyturia became detectable during normal ageing (up to 12 months). Despite podocyte detachment of varying degrees, no decrease in glomerular podocyte counts (i.e. WT-1-positive nuclei) was noted in either disease model.

The above observations suggest that the detection of proteinuria cannot distinguish between a persistent defect of the glomerular barrier and ongoing injury, whereas podocyturia may be more specific for 'active' injury.

Podocyturia in the PAN and anti-Thy 1.1 nephritis model was preceded by entry of glomerular podocytes into the cell cycle, with cyclin D1, cdc2 and/or PCNA expression [29]. Further, the peak of podocytic cell cycle activity coincided with the peak of podocyturia, which extends and confirms previous data in PHN [30]. These observations argue against the hypothesis that cell cycle entry is merely a response to detachment and urinary loss of podocytes. Rather they suggest that the attempt of podocytes to enter cell division is associated with altered attachment to the GBM with at least in some cases subsequent loss into the urinary space.

Detachment of Viable Podocytes in Patients with Glomerular Diseases

The occurrence of viable podocytes in human urine was also sought in cytospins stained with a podocalyxin antibody (with all the caveats about specificity of podocalyxin as discussed above). Patients with active FSGS and lupus nephritis indeed exhibited podocyturia, whereas those with inactive disease and normal controls did not [31]. Notably, in the positive patients the extent of podocyturia and proteinuria did not correlate, which supports our above hypothesis of podocyturia and proteinuria indicating potentially different glomerular events. TUNEL staining revealed the presence of viable and apoptotic podocytes in the urine and 10–50% of the podocytes were bi- or multinucleated. Between 30 and 40% of the podocalyxin-positive cells were also positive for synaptopodin, GLEPP1 or podocin, other podocyte markers, whereas all were negative for WT-1 [31]. As in our experimental studies, the human podocytes cultured from urine became apoptotic after about 2 weeks.

Technical Issues in the Detection and Quantification of Podocyturia

So far all methods to detect podocyte detachment were semiquantitative, time consuming to perform and usually required freshly voided urine samples. All clinical studies so far relied on cytospins of fresh urine sediments followed by immunostaining with antibodies against podocalyxin [31, 32]. The alternative approach of ex vivo culture of urinary podocytes [30, 31] is unlikely to become a clinical diagnostic tool, since it is time consuming, has the problem of potential bacterial or fungal contamination especially in female patients and counting of cultivated cells is complicated by possible proliferation and/or apoptosis of podocytes in cell culture. Furthermore, this method will only detect cells that are able to attach to the culture dish and thus is likely to miss a considerable percentage of viable podocytes.

An attractive method to quantify podocyturia and to avoid many of the above problems is flow cytometry. Such a cytometric approach should be based on the detection of single or ideally multiple specific podocyte surface or intracellular antigens. By setting a cut-off for the DNA content and particle size, podocyte apoptotic cells and membrane vesicles that are shed into the urine from injured podocytes could be excluded from detection. Unfortunately, neither our own group nor that of Lemley [pers. commun.] so far has managed to get a flow cytometry-based approach to work reliably in human urine samples.

One likely limitation of all the above techniques is that it relies on podocytes maintaining their expression of specific proteins both in the glomerulus and following shedding. Whereas at least in situ podocytes maintain specific expression in most glomerular diseases, podocyte dedifferentiation is well recognized in conditions such as HIV-associated nephritis, collapsing FSGS or within crescents during rapidly progressive glomerulonephritis. In these instances, detection of podocyturia might be very difficult or impossible. Conversely, another level of complexity is added by the very recent description of parietal podocytes in human kidneys, i.e. podocytes on the outer leaflet of Bowman's capsule that maintain expression of podocyte specific proteins such as podocalyxin, podocin, nephrin, and synaptopodin [33]. So far, relative numbers of such podocytes as well as their (patho-)physiologic behavior, in particular responses to injury, are unknown.

Another interesting approach is the detection of urinary podocyte specific mRNA by RT-PCR, e.g. nephrin or podocin (reviewed in [18]). Isolation of mRNA from the urinary sediment needs fresh urine samples and is likely vulnerable to the presence of bacteria, proteases or RNAses. Another potential concern with this approach is that one can exclude podocytes that are entering apoptosis or necrosis. These later cells in turn might exhibit highly variable protein and mRNA contents depending on the stage and extent of apoptosis/necrosis. The clinical usefulness of detecting podocyte specific mRNA in the urinary sediment is currently unknown, in particular, since very recent reports failed to detect nephrin or podocalyxin mRNA in samples of 70 highly proteinuric patients [27].

In view of the above problems, various alternative non-invasive approaches to detect podocyte damage in vivo have been tested, including the quantification of urinary slit diaphragm molecules such as nephrin by immunoblotting [34–36] or of podocalyxin by ELISA [37]. None of these assays has reached clinical practice so far and further studies will be necessary before the value of these approaches can be judged.

In summary, experimental data support the notion that podocyturia is both an important contributor to glomerular damage and a potentially valuable diagnostic tool, in particular given its potential to detect ongoing glomerular injury as opposed to proteinuria, which may derive from both ongoing injury and residual damage after injury has already subsided. However, technical problems so far have prevented a broad application of podocyturia measurements in clinical practice.

Acknowledgements

This work was supported by grants from the RWTH University, Aachen, Germany (START program) and the Deutsche Forschungsgemeinschaft (SFB 542, C7).

References

1 Pagtalunan ME, Miller PL, Jumping-Eagle S, et al: Podocyte loss and progressive glomerular injury in type II diabetes. J Clin Invest 1997;99:342–348.

2 Steffes MW, Schmidt D, McCrery R, Basgen JM: Glomerular cell number in normal subjects and in type 1 diabetic patients. Kidney Int 2001;59:2104–2113.

3 Lemley KV, Lafayette RA, Safai M, et al: Podocytopenia and disease severity in IgA nephropathy. Kidney Int 2002;61:1475–1485.

4 White KE, Bilous RW, Marshall SM, et al: Podocyte number in normotensive type 1 diabetic patients with albuminuria. Diabetes 2002;51:3083–3089.

5 Dalla Vestra M, Masiero A, Roiter AM, Saller A, Crepaldi G, Fioretto P: Is podocyte injury relevant in diabetic nephropathy? Studies in patients with type 2 diabetes. Diabetes 2003;52:1031–1035.

6 Chen CA, Hwang JC, Guh JY, Chang JM, Lai YH, Chen HC: Reduced podocyte expression of alpha3beta1 integrins and podocyte depletion in patients with primary focal segmental glomerulosclerosis and chronic PAN-treated rats. J Lab Clin Med 2006;147:74–82.

7 Kriz W: The pathogenesis of 'classic' focal segmental glomerulosclerosis – lessons from rat models. Nephrol Dial Transplant 2003;18(suppl 6):vi39–vi44.

8 Petermann A, Hiromura K, Pippin J, et al: Differential expression of d-type cyclins in podocytes in vitro and in vivo. Am J Pathol 2004;164:1417–1424.

9 Wiggins JE, Goyal M, Sanden SK, et al: Podocyte hypertrophy, 'adaptation,' and 'decompensation' associated with glomerular enlargement and glomerulosclerosis in the aging rat: prevention by calorie restriction. J Am Soc Nephrol 2005;16:2953–2966.

10 Petermann AT, Pippin J, Durvasula R, et al: Mechanical stretch induces podocyte hypertrophy in vitro. Kidney Int 2005;67:157–166.

11 Petermann AT, Pippin J, Hiromura K, et al: Mitotic cell cycle proteins increase in podocytes despite lack of proliferation. Kidney Int 2003;63:113–122.

12 Floege J, Johnson RJ, Alpers CE, et al: Visceral glomerular epithelial cells can proliferate in vivo and synthesize platelet-derived growth factor β-chain. Am J Pathol 1993;142:637–650.

13 Schiffer M, Bitzer M, Roberts IS, et al: Apoptosis in podocytes induced by TGF-beta and Smad7. J Clin Invest 2001;108:807–816.

14 Durvasula RV, Petermann AT, Hiromura K, et al: Activation of a local tissue angiotensin system in podocytes by mechanical strain. Kidney Int 2004;65:30–39.

15 Schiffer M, Mundel P, Shaw AS, Bottinger EP: A novel role for the adaptor molecule CD2-associated protein in transforming growth factor-beta-induced apoptosis. J Biol Chem 2004;279:37004–37012.

16 Huber TB, Hartleben B, Kim J, et al: Nephrin and CD2AP associate with phosphoinositide 3-OH kinase and stimulate AKT-dependent signaling. Mol Cell Biol 2003;23:4917–4928.

17 Bijian K, Takano T, Papillon J, Khadir A, Cybulsky AV: Extracellular matrix regulates glomerular epithelial cell survival and proliferation. Am J Physiol 2004;286:F255–F266.

18 Colucci G, Floege J, Schena FP: The urinary sediment beyond light microscopical examination. Nephrol Dial Transplant 2006;21:1482–1485.

19 Hara M, Yanagihara T, Takada T, et al: Urinary excretion of podocytes reflects disease activity in children with glomerulonephritis. Am J Nephrol 1998;18:35–41.

20 Hara M, Yanagihara T, Kihara I: Urinary podocytes in primary focal segmental glomerulosclerosis. Nephron 2001;89:342–347.

21 Nakamura T, Ushiyama C, Suzuki S, et al: Urinary excretion of podocytes in patients with diabetic nephropathy. Nephrol Dial Transplant 2000;15:1379–1383.

22 Nakamura T, Ushiyama C, Osada S, Hara M, Shimada N, Koide H: Pioglitazone reduces urinary podocyte excretion in type 2 diabetes patients with microalbuminuria. Metabolism 2001;50:1193–1196.

23 Nakamura T, Ushiyama C, Shimada N, et al: Effect of the antiplatelet drug dilazep dihydrochloride on urinary podocytes in patients in the early stage of diabetic nephropathy. Diabetes Care 2000;23:1168–1171.

24 Nakamura T, Ushiyama C, Hirokawa K, et al: Effect of cerivastatin on proteinuria and urinary podocytes in patients with chronic glomerulonephritis. Nephrol Dial Transplant 2002;17:798–802.

25 Nakamura T, Ushiyama C, Suzuki S, et al: Effects of angiotensin-converting enzyme inhibitor, angiotensin II receptor antagonist and calcium antagonist on urinary podocytes in patients with IgA nephropathy. Am J Nephrol 2000;20:373–379.

26 Hara M, Yanagihara T, Kihara I, Higashi K, Fujimoto K, Kajita T: Apical cell membranes are shed into urine from injured podocytes: a novel phenomenon of podocyte injury. J Am Soc Nephrol 2005;16:408–416.
27 Ihmoda I, Saleem M, Mathieson P, Turner P: Podocyturia: a useful test for progressive renal diseases (abstract)? Nephrol Dial Transpl 2006;21(suppl 4):323.
28 Petermann AT, Pippin J, Krofft R, et al: Viable podocytes detach in experimental diabetic nephropathy: potential mechanism underlying glomerulosclerosis. Nephron Exp Nephrol 2004;98:e114–e123.
29 Yu D, Petermann A, Kunter U, Rong S, Shankland SJ, Floege J: Urinary podocyte loss is a more specific marker of ongoing glomerular damage than proteinuria. J Am Soc Nephrol 2005;16:1733–1741.
30 Petermann AT, Krofft R, Blonski M, et al: Podocytes that detach in experimental membranous nephropathy are viable. Kidney Int 2003;64:1222–1231.
31 Vogelmann SU, Nelson WJ, Myers BD, Lemley KV: Urinary excretion of viable podocytes in health and renal disease. Am J Physiol 2003;285:F40–F48.
32 Hara M, Yanagihar T, Itoh M, Matsuno M, Kihara I: Immunohistochemical and urinary markers of podocyte injury. Pediatr Nephrol 1997;12:43–48.
33 Bariety J, Mandet C, Hill GS, Bruneval P: Parietal podocytes in normal human glomeruli. J Am Soc Nephrol 2006;17:2770–2780.
34 Luimula P, Aaltonen P, Ahola H, Palmen T, Holthofer H: Alternatively spliced nephrin in experimental glomerular disease of the rat. Pediatr Res 2000;48:759–762.
35 Kelly DJ, Aaltonen P, Cox AJ, et al: Expression of the slit-diaphragm protein, nephrin, in experimental diabetic nephropathy: differing effects of anti-proteinuric therapies. Nephrol Dial Transplant 2002;17:1327–1332.
36 Patari A, Forsblom C, Havana M, Taipale H, Groop PH, Holthofer H: Nephrinuria in diabetic nephropathy of type 1 diabetes. Diabetes 2003;52:2969–2974.
37 Kanno K, Kawachi H, Uchida Y, Hara M, Shimizu F, Uchiyama M: Urinary sediment podocalyxin in children with glomerular diseases. Nephron Clin Pract 2003;95:c91–c99.

Note Added in Print

Cumulative podocyturia was recently identified as a highly sensitive and relatively specific marker for disease progression in children and young adults with IgA nephropathy and Schönlein-Henoch purpura [1].

Reference

1 Hara M, Yanagihara T, Kihara I: Cumulative excretion of urinary podocytes reflects disease progression in IgA nephropathy and Schönlein-Henoch purpura. C J Am Soc Nephrol 2007;2:231–238.

nephron
Clinical Practice

Nephron Clin Pract 2007;106:c67–c71
DOI: 10.1159/000101800

Published online: June 6, 2007

Podocytes in HIV-Associated Nephropathy

Ting-Chi Lu John Cijiang He Paul E. Klotman

Division of Nephrology, Department of Medicine, Mount Sinai School of Medicine, New York, N.Y., USA

Key Words
HIV-associated nephropathy · Podocyte · Protein expression · Retinoids

Abstract
HIV-associated nephropathy (HIVAN) is the leading cause of end-stage renal failure in HIV-1 seropositive patients. The pathologic findings include collapsing focal segmental glomerulosclerosis with proliferation of epithelial cells in Bowman's space. Anatomically, these cells correspond to podocytes and exhibit a unique phenotype with loss of many differentiation markers including synaptopodin and dysregulation of the cell cycle markers consistent with proliferation. Podocyte dysfunction appears to be a direct result of HIV-1 protein expression, specifically Nef and Vpr as well as specific host factors that have yet to be elucidated. The mechanism by which Nef induces podocyte proliferation and dedifferentiation has been traced to its ability to activate several signaling pathways including Src-Stat3 and ras-raf-MAPK1, 2. Activation of the cAMP/PKA pathway with all-trans-retinoic acid appears to modulate these changes and returns podocytes to a differentiated, nonproliferating phenotype.

Introduction

HIV-associated nephropathy (HIVAN) is the most common cause of end-stage renal disease in patients with HIV-1 infection [1]. Affected patients generally have advanced HIV-1 infection and present clinically with nephrotic range proteinuria and progressive azotemia [2–4]. Since its first description in the mid 1980s, much has been learned about the unique pathogenesis of this disease. Although HIVAN involves all compartments of the kidney, podocytes are one of the primary cell types affected, and nephropathy is due primarily to HIV-1 infection of renal epithelium and expression of HIV-1 proteins. This review will focus on the changes HIV-1 infection brings about in podocyte phenotype and the molecular mechanisms by which specific viral proteins induce these changes.

Pathological Findings of Podocyte Injury

Renal biopsies from patients with HIVAN display collapsing focal segmental glomerulosclerosis, interstitial nephritis, and tubular microcystic dilatation [5, 6]. A distinctive feature of this disease is the finding of prominent hypertrophy and hyperplasia of the epithelial cells overlying the glomerulus. These cells may also exhibit enlarged vesicular nuclei, mitotic figures and intracytoplasmic protein resorption droplets [7].

KARGER

Fax +41 61 306 12 34
E-Mail karger@karger.ch
www.karger.com

1660–2110/07/1062–0067$23.50/0

Accessible online at:
www.karger.com/nec

Ting-Chi Lu
One Gustave L. Levy Place
Box 1243
New York, NY 10029 (USA)
Tel. +1 212 241 1423, Fax +1 212 987 0389, E-Mail Ting-chi.Lu@mssm.edu

Immunohistochemical analyses of biopsies from patients with HIVAN are remarkable for decreased staining for CALLA, C3b receptor, GLEPP-1, podocalyxin, synaptopodin, and WT-1 in areas of glomerulosclerosis. These are all markers normally seen in mature podocytes. In contrast, other glomerular diseases that are associated with nephrotic syndrome such as membranous nephropathy and minimal change disease retain the mature podocyte phenotype [8]. The loss of synaptopodin staining appears to be especially specific for HIVAN and is frequently found even in unaffected areas of the glomerulus [8].

Cell cycle regulation is also disrupted in HIVAN. Normal, mature podocytes express cyclin D1 and the cyclin kinase inhibitors (CKI) p21and p27 consistent with a quiescent cell [9]. In HIVAN, however, these proteins are decreased. Conversely, staining for Ki-67 and cyclin A is increased in podocytes within sclerotic areas of the glomerulus [10]. These data suggest that unlike other glomerular diseases, HIV-1 infection leads to podocyte proliferation and de-differentiation.

The identity of the proliferating cells in Bowman's space has been a recent area of interest. Whether these cells represent podocytes or proliferating parietal epithelial cells has been an area of debate. In the early stages of disease, the proliferating cells remain attached to the glomerular basement membrane (GBM) suggesting anatomically that they are podocytes [8]. A recent study used immunohistochemistry for a variety of podocyte-specific and parietal epithelial cell markers to stain biopsies obtained from two patients with HIVAN. Though the study showed that the proliferating cells did not stain positively as podocytes [11], the results are inconclusive since de-differentiation and loss of podocyte-specific markers is central to the pathogenesis of HIV nephropathy [8]. Further cell lineage studies are needed to clarify this issue, but at this time, there is a great deal of in vivo and in vitro data suggesting that because of the unique pathogenesis of HIV nephropathy, podocytes in this disease tend to proliferate.

The experimental model of HIVAN, the Tg26 mouse model, recapitulates many of the findings seen in humans. Glomerular staining for synaptopodin is markedly decreased whereas staining for Ki-67 and desmin is increased [12]. Moreover, conditionally immortalized podocyte lines derived from Tg26 mice display increased growth rate, loss of contact inhibition, cell cycle progression, and decreased expression of WT-1, synaptopodin, podocalyxin when compared to control podocytes [13]. Thus, the Tg26 mouse model is an invaluable tool for studying podocyte pathology.

Viral Infection of Podocytes

Two lines of evidence support the hypothesis that the podocyte phenotype is a direct result of HIV-1 infection of the cells and not a reaction to a systemic HIV-1 infection or mediated by immune-complex deposition. HIV-1 can be detected in human biopsies and transgenic animal models of HIVAN develop disease phenotypically identical to HIVAN when the transgene is expressed in podocytes. In man, using in situ hybridization to examine biopsies from HIV-1 seropositive patients, HIV-1 RNA was detected in renal epithelial cells. The presence of HIV-1 was further confirmed using in situ polymerase chain reaction (PCR) of DNA. Both HIV-1 RNA and DNA were detectable in renal epithelia from several patients with undetectable viral loads, suggesting that renal cells may act as a reservoir for HIV-1 [14, 15]. Additional evidence for a renal reservoir of HIV-1 was provided by a case report of a patient who developed HIVAN with acute HIV-1 infection. Treatment with highly active antiretroviral therapy resulted in resolution of clinical and pathologic signs of the disease. Renal epithelial expression of viral mRNA, however, did not change following treatment [16]. Phylogenetic analysis of gp120 DNA sequences obtained from infected renal tubular cells and peripheral blood mononuclear cells from the same patients revealed evidence of tissue-specific viral evolution, suggesting that renal epithelial cells are able to support full viral replication [17]. The mechanism of viral entry into renal epithelial cells, which lack CD4, is still unknown. CCR5 and CXCR4 are coreceptors for CD4 and mediate HIV-1 entry into lymphocytes. These receptors have not been demonstrated on renal epithelial cells in vivo [18].

Animal studies also confirm a direct effect of HIV expression on the renal phenotype. In the murine model of HIVAN, expression of HIV-1 mRNA is detectable in renal epithelium [12]. Reciprocal kidney transplants between wild-type and Tg26 mice demonstrate that HIVAN develops in transgenic kidneys transplanted into nontransgenic littermates while normal kidneys remain disease-free when transplanted into transgenic littermates [19]. Further, when the *gag-pol* deleted HIV-1 genome was selectively expressed in the podocytes of transgenic mice using the *Nphs1* promoter, mice not only developed massive proteinuria and severe collapsing FSGS, but also microcystic tubular dilatations [20]. This study highlights the central role podocyte infection plays in HIVAN pathogenesis.

Podocyte Response to Viral Infection

HIVAN affects patients of African descent almost exclusively [2, 21, 22]. In the experimental setting FVB/N mice are susceptible to development of proteinuria and the histopathologic changes consistent with HIVAN in the presence of the HIV-1 transgene expression, whereas strains such as C57BL/6 are resistant [23–25]. These findings highlight the importance of host genetic factors in the pathogenesis of HIVAN. To delineate these factors, representational difference analysis (RDA) of cDNA obtained from HIV-1 podocytes was compared to wild-type podocytes. Two novel proteins, sidekick-1 (Sdk-1) and podocan were identified by this method [26, 27].

Sdk-1 is a transmembrane protein of the Ig superfamily. Normally, it is not identified in adult podocytes; however, in the setting of HIV-1 transgene expression, the protein is highly upregulated [26]. Sdk-1 and its ortholog, sidekick-2 (Sdk-2) function as homophilic binding proteins that regulate cell-cell adhesion [28]. Data suggest that sidekicks may mediate podocyte clumping in Bowman's space. Studies are ongoing to determine their precise role in HIVAN pathogenesis.

Podocan is a member of the small leucine-rich repeat family of proteins. Two transcripts are detectable in normal kidneys, a 3.2 kb transcript during embryogenesis and a 2.3-kb transcript that becomes expressed after birth. The 3.2-kb transcript encodes a 611 amino acid secretory protein with 20 leucine-rich repeats. In situ hybridization of normal kidney reveals podocan mRNA in podocytes and also in glomerular endothelial cells. Immunohistochemical staining in normal glomeruli localizes it to the glomerular basement membrane. Staining for podocan was markedly increased in sclerotic glomerular lesions of HIVAN [27]. The functional role of podocan and its relationship to HIVAN pathogenesis remains to be determined.

Role of Specific Viral Proteins in Podocyte Dysfunction

In vitro studies have identified Nef, an HIV-1 accessory protein, as one of the primary mediators of the changes in podocyte phenotype in HIVAN. Infection of murine podocytes with an HIV-1 proviral construct that expresses *nef* causes proliferation, anchorage-independent growth, loss of podocyte differentiation markers and increases in podocyte proliferation markers. On the other hand, infection of murine podocytes with full-length HIV-1 containing mutated *nef* fails to recapitulate the phenotypic changes found in HIVAN [29, 30].

Several additional murine models have been developed to identify the viral proteins responsible for HIVAN pathogenesis. This work has confirmed the importance of Nef in mediating HIVAN. A mouse model with podocyte specific expression of Nef developed evidence of cell cycle disruption in podocytes with increased staining for Ki-67 as well as downregulation of the podocyte differentiation markers synaptopodin and WT-1. Furthermore, increased staining for phospho-Stat3 was found in the nef-expressing podocytes confirming previous in vitro data discussed below [31].

A recent study carefully evaluated the importance of individual HIV gene expression in the podocytes of transgenic mice. In this study, expression of either *vpr* or *nef* alone was sufficient to induce development of podocyte injury, as evidenced by decreased synaptopodin and nephrin expression and glomerulosclerosis. Conversely, mice expressing *vif, tat, rev,* and *vpu* were not affected. Furthermore, mice that expressed both *nef* and *vpr* developed severe podocyte injury and glomerulosclerosis suggesting that the two viral genes play a synergistic role in inducing podocyte damage and glomerulopathy [32].

HIV-Viral Protein-Induced Cell Signaling in Podocytes

Nef is a 27-kDa, myristoylated protein that plays a key role in HIV-1 pathogenicity by enhancing viral infectivity and replication [33–36]. Nef has multiple functional domains that allow it to interact with several cell-signaling pathways, depending on its intracellular location [36–39]. The proline-rich (PxxP) domain, which allows it to activate signaling proteins such as Src family kinases, is of key importance to HIVAN pathogenesis.

Src family kinases are nonreceptor tyrosine kinases that play vital roles in regulating cell proliferation, cell-cell adhesion and cell motility [40]. The Src SH3 domain allows for a critical interaction with Nef. This leads to ras-c-raf-MAPk1, 2 activation as well as Stat3 activation which in turn leads to podocyte proliferation and dedifferentiation [41]. As mentioned before, the PxxP domain of Nef is critical in this signal cascade: replacement of wild-type Nef with *nef* that has a mutation in the PxxP region abolishes the Nef-induced podocyte proliferation. Src acts upstream of both pathways since transfecting HIV-1 podocytes with a dominant negative form of Src inhibits activation of Stat3 as well as MAPk1, 2. The dominant negative Src also prevents HIV infection from inducing dedifferentiation and proliferation of podocytes [41]. The importance of these two signaling pathways in podocytes has been confirmed in vivo. Biopsies from patients with HIVAN were shown to have increased staining for phosphorylated Stat3 and MAPK1, 2. This was in contrast to a normal kidney biopsy as well as non-HIV kidney disease (FSGS) [41]. A mouse model expressing Nef selectively in podocytes also develops increased staining for phospho-Stat3 in podocytes in conjunction with early markers of proliferation and dedifferentiation [31].

Podocyte Injury as a Target of Therapy for HIVAN

A recent study demonstrated that all-trans-retinoic acid (atRA) is able to redirect cell signaling in podocytes to reverse the changes induced by Nef. Retinoids are derivatives of vitamin A and have multiple cellular functions including inhibition of proliferation, induction of cell differentiation, regulation of apoptosis, and inhibition of inflammation [42]. In addition to their established benefits in treating various malignancies, retinoids have been found to provide protection in several experimental models of kidney disease [43–45]. Retinoids classically exert their effects on cells by binding to retinoic acid receptors (RAR) in the cytoplasm which in turn bind to retinoic acid response elements (RARE) in the promoter regions of genes. However, induction of cytosolic signaling molecules by retinoids has also been reported [46]. Treatment of HIV-1-infected podocytes with retinoic acid results in inhibition of proliferation by cell cycle arrest at G_1. Retinoids also restore the expression of podocyte differentiation markers synaptopodin, WT-1, nephrin and podocin in HIV-1 infected podocytes. In vivo treatment of Tg26 mice with atRA results in decreased proteinuria. Retinoic acid induces this reversal in podocyte phenotype by activating the cAMP/protein kinase A (PKA) pathway. This signaling cascade culminates in activation of the transcription factor CREB (cAMP response element binding protein) and conversion of MAPK1, 2 activation from a sustained to a transient pattern [47]. These findings highlight a potential role for treatment specifically targeted at podocytes with retinoic acid.

Conclusion and Future Directions

The unique phenotype induced by HIV-1 infection has provided valuable insight into podocyte physiology and pathology. In particular, studies of the molecular mechanisms by which HIV induces podocyte proliferation has elucidated many of the mechanisms controlling cell cycle and signaling that are crucial to maintaining a differentiated podocyte phenotype. Future studies that address the mechanisms of altered cytoskeleton induced by HIV infection should provide similar insights. Another area of importance is the mechanism by which HIV-1 is able to infect podocytes. Finally, our study of atRA suggests that ameliorating podocyte injury could be a major adjunctive therapy for HIVAN and any disease in which podocytes proliferate. Answers to the remaining questions should have broad implications for our understanding of not only HIVAN pathogenesis but glomerular diseases in general.

References

1 USRDS 2005 Annual Data Report: Atlas of End-Stage-Renal-Disease in the United States. Bethesda, National Institutes of Health, National Institute of Diabetes and Digestive and Kidney Diseases, 2005/2006.

2 Rao TK, Filippone EJ, Nicastri AD, et al: Associated focal and segmental glomerulosclerosis in the acquired immunodeficiency syndrome. N Engl J Med 1984;310:669–673.

3 Rao TK, Friedman EA, Nicastri AD: The types of renal disease in the acquired immunodeficiency syndrome. N Engl J Med 1987; 316:1062–1068.

4 Carbone L, D'Agati V, Cheng JT, Appel GB: Course and prognosis of human immunodeficiency virus-associated nephropathy. Am J Med 1989;87:389–395.

5 D'Agati V, Suh JI, Carbone L, Cheng JT, Appel G: Pathology of HIV-associated nephropathy: a detailed morphologic and comparative study. Kidney Int 1989;35:1358–1370.

6 Chander P, Soni A, Suri A, Bhagwat R, Yoo J, Treser G: Renal ultrastructural markers in AIDS-associated nephropathy. Am J Pathol 1987;126:513–526.

7 D'Agati V, Appel GB: HIV infection and the kidney. J Am Soc Nephrol 1997;8:138–152.

8 Barisoni L, Kriz W, Mundel P, D'Agati V: The dysregulated podocyte phenotype: a novel concept in the pathogenesis of collapsing idiopathic focal segmental glomerulosclerosis and HIV-associated nephropathy. J Am Soc Nephrol 1999;10:51–61.

9 Nagata M, Nakayama K, Terada Y, Hoshi S, Watanabe T: Cell cycle regulation and differentiation in the human podocyte lineage. Am J Pathol 1998;153:1511–1520.

10 Barisoni L, Mokrzycki M, Sablay L, Nagata M, Yamase H, Mundel P: Podocyte cell cycle regulation and proliferation in collapsing glomerulopathies. Kidney Int 2000;58:137–143.

11 Dijkman HB, Weening JJ, Smeets B, et al: Proliferating cells in HIV and pamidronate-associated collapsing focal segmental glomerulosclerosis are parietal epithelial cells. Kidney Int 2006;70:338–344.
12 Barisoni L, Bruggeman LA, Mundel P, D'Agati VD, Klotman PE: HIV-1 induces renal epithelial dedifferentiation in a transgenic model of HIV-associated nephropathy. Kidney Int 2000;58:173–181.
13 Schwartz EJ, Cara A, Snoeck H, et al: Human immunodeficiency virus-1 induces loss of contact inhibition in podocytes. J Am Soc Nephrol 2001;12:1677–1684.
14 Cohen AH, Sun NC, Shapshak P, Imagawa DT: Demonstration of human immunodeficiency virus in renal epithelium in HIV-associated nephropathy. Mod Pathol 1989;2:125–128.
15 Bruggeman LA, Ross MD, Tanji N, et al: Renal epithelium is a previously unrecognized site of HIV-1 infection. J Am Soc Nephrol 2000;11:2079–2087.
16 Winston JA, Bruggeman LA, Ross MD, et al: Nephropathy and establishment of a renal reservoir of HIV type 1 during primary infection. N Engl J Med 2001;344:1979–1984.
17 Marras D, Bruggeman LA, Gao F, et al: Replication and compartmentalization of HIV-1 in kidney epithelium of patients with HIV-associated nephropathy. Nat Med 2002;8:522–526.
18 Shankland SJ, Eitner F, Hudkins KL, Goodpaster T, D'Agati V, Alpers CE: Differential expression of cyclin-dependent kinase inhibitors in human glomerular disease: role in podocyte proliferation and maturation. Kidney Int 2000;58:674–683.
19 Bruggeman LA, Dikman S, Meng C, Quaggin SE, Coffman TM, Klotman PE: Nephropathy in human immunodeficiency virus-1 transgenic mice is due to renal transgene expression. J Clin Invest 1997;100:84–92.
20 Zhong J, Zuo Y, Ma J, et al: Expression of HIV-1 genes in podocytes alone can lead to the full spectrum of HIV-1-associated nephropathy. Kidney Int 2005;68:1048–1060.
21 Abbott KC, Hypolite I, Welch PG, Agodoa LY: Human immunodeficiency virus/acquired immunodeficiency syndrome-associated nephropathy at end-stage renal disease in the United States: patient characteristics and survival in the pre highly active antiretroviral therapy era. J Nephrol 2001;14:377–383.
22 Laradi A, Mallet A, Beaufils H, Allouache M, Martinez F: HIV-associated nephropathy: outcome and prognosis factors. Groupe d' Etudes Nephrologiques d'Ile de France. J Am Soc Nephrol 1998;9:2327–2335.
23 Dickie P, Felser J, Eckhaus M, et al: HIV-associated nephropathy in transgenic mice expressing HIV-1 genes. Virology 1991;185:109–119.
24 Kopp JB, Klotman ME, Adler SH, et al: Progressive glomerulosclerosis and enhanced renal accumulation of basement membrane components in mice transgenic for human immunodeficiency virus type 1 genes. Proc Natl Acad Sci USA 1992;89:1577–1581.
25 Gharavi AG, Ahmad T, Wong RD, et al: Mapping a locus for susceptibility to HIV-1-associated nephropathy to mouse chromosome 3. Proc Natl Acad Sci USA 2004;101:2488–2493.
26 Kaufman L, Hayashi K, Ross MJ, Ross MD, Klotman PE: Sidekick-1 is upregulated in glomeruli in HIV-associated nephropathy. J Am Soc Nephrol 2004;15:1721–1730.
27 Ross MD, Bruggeman LA, Hanss B, et al: Podocan, a novel small leucine-rich repeat protein expressed in the sclerotic glomerular lesion of experimental HIV-associated nephropathy. J Biol Chem 2003;278:33248–33255.
28 Hayashi K, Kaufman L, Ross MD, Klotman PE: Definition of the critical domains required for homophilic targeting of mouse sidekick molecules. FASEB J 2005;19:614–616.
29 Husain M, Gusella GL, Klotman ME, et al: HIV-1 Nef induces proliferation and anchorage-independent growth in podocytes. J Am Soc Nephrol 2002;13:1806–1815.
30 Sunamoto M, Husain M, He JC, Schwartz EJ, Klotman PE: Critical role for Nef in HIV-1-induced podocyte dedifferentiation. Kidney Int 2003;64:1695–1701.
31 Husain M, D'Agati VD, He JC, Klotman ME, Klotman PE: HIV-1 Nef induces dedifferentiation of podocytes in vivo: a characteristic feature of HIVAN. AIDS 2005;19:1975–1980.
32 Zuo Y, Matsusaka T, Zhong J, et al: HIV-1 genes vpr and nef synergistically damage podocytes, leading to glomerulosclerosis. J Am Soc Nephrol 2006;17:2832–2843.
33 Kestler HW 3rd, Ringler DJ, Mori K, et al: Importance of the nef gene for maintenance of high virus loads and for development of AIDS. Cell 1991;65:651–662.
34 Fackler OT, Baur AS: Live and let die: Nef functions beyond HIV replication. Immunity 2002;16:493–497.
35 Papkalla A, Munch J, Otto C, Kirchhoff F: Nef enhances human immunodeficiency virus type 1 infectivity and replication independently of viral coreceptor tropism. J Virol 2002;76:8455–8459.
36 Baur AS, Sawai ET, Dazin P, Fantl WJ, Cheng-Mayer C, Peterlin BM: HIV-1 Nef leads to inhibition or activation of T cells depending on its intracellular localization. Immunity 1994;1:373–384.
37 Federico M, Percario Z, Olivetta E, et al: HIV-1 Nef activates STAT1 in human monocytes/macrophages through the release of soluble factors. Blood 2001;98:2752–2761.
38 Messmer D, Bromberg J, Devgan G, Jacque JM, Granelli-Piperno A, Pope M: Human immunodeficiency virus type 1 Nef mediates activation of STAT3 in immature dendritic cells. AIDS Res Hum Retrovir 2002;18:1043–1050.
39 Tobiume M, Fujinaga K, Suzuki S, Komoto S, Mukai T, Ikuta K: Extracellular Nef protein activates signal transduction pathway from Ras to mitogen-activated protein kinase cascades that leads to activation of human immunodeficiency virus from latency. AIDS Res Hum Retrovir 2002;18:461–467.
40 Martin GS: The hunting of the Src. Nat Rev Mol Cell Biol 2001;2:467–475.
41 He JC, Husain M, Sunamoto M, et al: Nef stimulates proliferation of glomerular podocytes through activation of Src-dependent Stat3 and MAPK1, 2 pathways. J Clin Invest 2004;114:643–651.
42 Evans TR, Kaye SB: Retinoids: present role and future potential. Br J Cancer 1999;80:1–8.
43 Lehrke I, Schaier M, Schade K, et al: Retinoid receptor-specific agonists alleviate experimental glomerulonephritis. Am J Physiol 2002;282:F741–F751.
44 Wagner J, Dechow C, Morath C, et al: Retinoic acid reduces glomerular injury in a rat model of glomerular damage. J Am Soc Nephrol 2000;11:1479–1487.
45 Moreno-Manzano V, Mampaso F, Sepulveda-Munoz JC, et al: Retinoids as a potential treatment for experimental puromycin-induced nephrosis. Br J Pharmacol 2003;139:823–831.
46 Licht JD, Zelent A: Retinoid and growth factor receptor signaling in APL. Blood 2005;105:1381–1382.
47 He JC, Lu TC, Fleet M, Sunamoto M, Husain M, Fang W, Neves S, Chen Y, Shankland S, Iyengar R, Klotman PE: Retinoic acid inhibits HIV-1 induced podocyte proliferation through the cAMP pathway. J Am Soc Nephrol 2007;18:93–102.

nephron
Clinical Practice

Nephron Clin Pract 2007;106:c72–c81
DOI: 10.1159/000101801

Published online: June 6, 2007

Pre-Eclampsia: Clinical Manifestations and Molecular Mechanisms

Suzanne Baumwell[a] S. Ananth Karumanchi[a, b]

Departments of [a]Medicine and [b]Obstetrics and Gynecology, Beth Israel Deaconess Medical Center and Harvard Medical School, Boston, Mass., USA

Key Words
Hypertension · Proteinuria · Angiogenesis · Vascular endothelial growth factor · Endotheliosis

Abstract
Preeclampsia affects 3–5% of pregnancies and can have a significant impact on health for both mother and fetus. Risk factors include maternal co-morbidities such as obesity and chronic hypertension, paternal factors, and genetic factors. New hypertension and proteinuria during the second half of pregnancy are key diagnostic criteria, but the clinical features and associated prognostic implications are somewhat heterogeneous and may reflect different mechanisms of disease. Renal dysfunction and proteinuria correspond to the pathologic finding of glomerular endotheliosis, and generally resolve after cure of preeclampsia through fetal and placenta delivery. The molecular mechanisms behind this disease are being discovered and refined. The initial etiologic agents are currently unknown. Pathologic studies show abnormal development of an ischemic placenta with a high-resistance vasculature, which cannot deliver an adequate blood supply to the fetoplacental unit. Endothelial dysfunction plays a central role in the pathogenesis of the maternal syndrome. Dysfunctional endothelial cells produce altered quantities of vasoactive mediators, which lead to a tip in the balance towards vasoconstriction. An imbalance in circulating angiogenic factors is emerging as a prominent mechanism that mediates the endothelial dysfunction and the clinical signs and symptoms of preeclampsia. Soluble fms-like tyrosine kinase 1 (sFlt1), an endogenous anti-angiogenic factor that is a potent vascular endothelial growth factor (VEGF) antagonist, is highly elevated in preeclampsia. VEGF is not only important in angiogenesis, but also in maintaining endothelial health including the formation of endothelial fenestrae (a hallmark of the glomerular vascular endothelium). sFlt1 overexpression in animals induces glomerular endotheliosis with the loss of endothelial fenestrae that resembles the renal histological lesions of preeclampsia. More severe forms of preeclampsia, including the HELLP syndrome, may be explained by a concomitant elevation in both sFlt1 and soluble endoglin, another anti-angiogenic factor. Unraveling of the molecular mechanisms behind preeclampsia may help to expand our armamentarium to treat patients in a more directed fashion, as current management consists of supportive care and expedited delivery. Finally, long-term outcomes of women with preeclampsia include a significantly increased risk for hypertension and cardiovascular disease, including mortality, which may warrant more aggressive screening and treatment in this population.

KARGER

Fax +41 61 306 12 34
E-Mail karger@karger.ch
www.karger.com

Accessible online at:
www.karger.com/nec

S. Ananth Karumanchi, MD
Beth Israel Deaconess Medical Center, Renal Division, RW 663
330 Brookline Avenue
Boston, MA 02215 (USA)
Tel. +1 617 667 1018, Fax +1 617 667 2913, E-Mail sananth@bidmc.harvard.edu

Introduction

Pre-eclampsia is a heterogeneous disorder affecting 3–5% of pregnancies and a major cause of obstetric morbidity and mortality worldwide. For the fetus, pre-eclampsia can result in small-for-gestational-age infancy, preterm delivery, hypoxic neurologic injury or death. For the mother, complications of pre-eclampsia include renal failure, HELLP syndrome (hemolysis, elevated liver enzymes, and thrombocytopenia), seizures, liver failure, stroke or death. Women with pre-eclampsia often later develop cardiovascular disease and hypertension, another source of significant morbidity.

While the initial events involved in the pathogenesis of pre-eclampsia are still a mystery, recent developments have begun to elucidate molecular mechanisms behind its manifestations. An abnormally developed, hypoxic placenta and endothelial dysfunction are important themes. Further, there seems to be a tip in the balance of angiogenesis versus antiangiogenesis, with favoring of the latter in pre-eclampsia. This review will discuss some of the mechanisms leading to the clinical manifestations of pre-eclampsia.

Diagnosis

Historically, there has been difficulty reaching a consensus on diagnostic criteria for pre-eclampsia. Current criteria mandate new onset of hypertension and proteinuria after 20 weeks' gestation, the details of which are reviewed elsewhere [1]. However, there are atypical presentations of pre-eclampsia that call for similar management, as maternal and fetal outcomes are comparable to cases meeting strict criteria for pre-eclampsia [2]. For example, women with hypertension and other associated signs and symptoms (abdominal pain, neurologic symptoms, thrombocytopenia, elevated transaminases), but without proteinuria, are more likely to deliver preterm infants than mild pre-eclamptics [3]. The timing of pre-eclampsia can be atypical, before 20 weeks or up to 4 weeks postpartum. Preterm pre-eclampsia (≤37 weeks) is generally more severe with greater likelihood of small-for-gestational age infants. The pathophysiology behind these variable presentations is poorly defined but may reflect differing mechanisms.

Epidemiology and Risk Factors

Epidemiologic studies have identified genetic mutations, paternal factors and maternal factors that confer an increased risk of pre-eclampsia [4]. Relevant to the pathophysiology of pre-eclampsia are some of the maternal systemic comorbidities that increase the risk of pre-eclampsia, including diabetes, hypertension, and hyperlipidemia, which are associated with chronic inflammation and endothelial dysfunction. Further, increased prepregnancy body mass index confers increased risk of pre-eclampsia [5], independent of hyperlipidemia and insulin resistance. Thrombophilias increase risk, particularly with inheritance of multiple types [6]. Interestingly, smoking, which may be hypothesized to increase risk given the previous discussion, is actually protective against pre-eclampsia [7], for unclear reasons.

Clinical Manifestations

Hypertension

Multiple mechanisms contribute to hypertension in pre-eclampsia, ranging from sympathetic tone and the renin-angiotensin system to altered production of endothelial-derived vasoactive mediators.

Women with pre-eclampsia have increased sympathetic tone as measured by number of bursts of postganglionic sympathetic nerves compared with normotensive pregnant and hypertensive nonpregnant women [8]. Women with pre-eclampsia demonstrate a heightened pressor response to norepinephrine [9], and are more sensitive to angiotensin II [10]. This sensitivity may be mediated by AT1-B2 heterodimerization (angiotensin II receptor type 1 and bradykinin B2 receptor) [11] or by an autoantibody against the AT1 receptor (AT1-AA), which facilitates interaction between angiotensin II and the AT1 receptor [12]. Angiotensin II plays multiple roles in the pathology of pre-eclampsia, including contributing to hypertension, increasing oxidative stress via production of superoxide anions, and activating platelets [13].

Injured, dysfunctional endothelial cells in pre-eclampsia produce altered quantities of circulating vasoactive mediators. For example, prostaglandin I2 (PGI2, or prostacyclin), an endothelial-derived vasodilator, is decreased in pre-eclampsia. There may be a circulating factor inhibiting its synthesis, as incubation of endothelial cells with pre-eclamptic serum results in decreased prostacyclin production [14]. Similarly, thromboxane A2, a vasoconstrictor produced by platelets, is elevated in women with pre-eclampsia. The ratio of TxA2/PGI2 is elevated after 16 weeks in pre-eclamptic women [15], which may be a marker of the vasoconstrictor versus vasodilator balance. Other alterations are seen in endothelin levels [16], a vasoconstrictor, and nitric oxide [17], a vasodilator.

Finally, the local control of vascular tone is affected by endothelial dysfunction. Doppler studies of the brachial artery demonstrated that endothelial-dependent flow-mediated dilatation of blood vessels was reduced up to three years after pre-eclamptic pregnancy [18]. The duration of this effect after resolution of pre-eclampsia suggests a lasting effect on local regulation of vascular tone.

Renal Dysfunction and Proteinuria

Glomerular filtration rate (GFR) and renal plasma flow (RPF) increase by 40–60% during the first trimester in normal pregnancy [19], resulting in a fall of serum blood urea nitrogen and creatinine. In pre-eclampsia, GFR and RPF are reduced by about 30–40% compared with normal pregnancies [20], resulting in a BUN and creatinine similar to nonpregnant women.

Proteinuria ranges from less than 1 g to nephrotic-range proteinuria. Interestingly, in most renal biopsies the podocyte foot processes, normally effaced in nephrotic conditions, are relatively preserved in pre-eclampsia [21]. It is thought that loss of size, and perhaps charge, selectivity in the glomerulus contributes to proteinuria [22].

Glomerular capillary endotheliosis is the classic renal lesion seen in pre-eclampsia [23, 24], though, notably, this lesion may be present in gestational hypertension or normal pregnancies, usually to a milder degree [25]. Under light microscopy, the glomeruli appear relatively large. Endothelial cells lining the capillaries are swollen and hypertrophied, leaving less space for erythrocytes and giving the glomeruli their bloodless appearance. Mesangial cells also appear swollen, with mesangial interposition in severe cases. Electron microscopy reveals subendothelial electron-dense fibrinoid and granular deposits, as well as loss of endothelial fenestrae.

Loss of fenestrae is significant in reducing GFR. Filtrate passes through the fenestrae in the endothelial cells lining the glomerular capillaries to enter into the urinary space. Lafayette et al. [20] showed that the space occupied by fenestrae in endothelial cells is reduced by about 70%, causing a significant increase in endothelial cell resistance, which causes reduction in GFR. In addition, the subendothelial deposits of fibrin or fibrinogen derivatives cause a longer extracellular pathway for the formation of filtrate, thereby also reducing GFR. As will be discussed further, vascular endothelial growth factor (VEGF) is necessary for fenestrations in endothelial cells [26] and loss of VEGF function in podocytes in murine models results in glomerular endotheliosis and proteinuria [27]. Thus, the renal manifestations of pre-eclampsia, with loss of fenestrae and glomerular hypofiltration, are at least in part caused by decreased VEGF signaling in the renal tissue.

Within 8 weeks of delivery, glomerular enlargement and endothelial swelling usually resolve, along with the hypertension and proteinuria. In some, however, proteinuria persists for months. Renal function generally returns to normal, but focal segmental glomerulosclerosis occurs in a small percentage [28]. Clinically, even these women regain normal renal function with preserved GFR and resolution of proteinuria.

Edema

Women with pre-eclampsia often note edema, particularly in the face, hands, and feet, with sudden weight gain. The edema of pre-eclampsia is similar to the 'overfill' edema of acute glomerulonephritis, in which the renal dysfunction itself is responsible for sodium and fluid retention.

The exact mechanism of edema formation in pre-eclampsia is not fully elucidated. Renin and aldosterone are decreased compared with normal pregnancy, in contrast to 'underfill' edema; however, pre-eclamptic women experience increased sensitivity to angiotensin II, and so quantitative measurements of these hormones may be misleading. Endothelial permeability may contribute, as suggested by the faster disappearance of albumin-bound Evan's blue dye in pre-eclamptic women compared with controls [29]. Hypoalbuminemia may contribute to edema, but is likely not the sole cause given that albumin repletion does not induce diuresis and resolution of edema [30].

Hyperuricemia

Increased uric acid concentrations in pre-eclampsia were first noticed in the early 20th century [31]. Often, this rise occurs prior to the onset of reduced GFR and proteinuria. Some studies have shown correlation between levels of uric acid and severity of disease for both the mother, with more severe proteinuria [32], renal pathology [33], and morbidity [34]; and for the fetus, particularly fetal demise [35]. However, a recent meta-analysis [36] challenges the utility of serum uric acid levels in predicting complications, finding a particularly poor correlation between uric acid levels and fetal stillbirth, intrauterine demise, and neonatal death.

In pre-eclamptic pregnancies with hyperuricemia, uric acid levels rise as early as 10 weeks' gestation and continue to rise until 48 h postpartum [37]. It is unclear whether elevated serum uric acid is an active player, or simply a marker, in the pathogenesis of pre-eclampsia. For instance, a rat model of hyperuricemia with intact renal function results in hypertension [38]. Studies have begun to forge connections between uric acid and vascular damage and inflammation [37], though evidence is controversial in predicting the risk of future cardiovascular disease.

Neurologic Abnormalities and Eclampsia

Women with pre-eclampsia can develop headache, blurred vision, and temporary loss of vision. These changes are thought to be due to the combination of impaired autoregulation from acute hypertension and endothelial dysfunction. The concept of impaired autoregulation is similar to that of encephalopathy in hypertensive emergency, in which the rapidity of onset precludes adaptations made by cerebral vasculature [39]. Cerebral edema, visual symptoms, and headache ensue.

However, the presence of endothelial dysfunction distinguishes pre-eclamptic from purely hypertensive encephalopathy. Endothelial dysfunction, with vessel leakiness and subsequent parenchymal edema may account for the lower blood pressures required to produce cerebral edema in pre-eclampsia compared with pure hypertensive encephalopathy [40]. Interestingly, one study found an association of brain edema on MRI with abnormal red blood cell morphology and increased LDH in pre-eclampsia, suggesting a link between microangiopathic hemolytic anemia and endothelial damage in cases of radiologically visible cerebral edema [39].

Women with pre-eclampsia are at risk of developing eclampsia, which is differentiated from the former by the presence of seizures. Often the onset of eclampsia is heralded by visual changes and severe headache. In fact, the origin of 'eclampsia' comes from the Greek word 'eklampsis', meaning 'shining forth,' which is thought to describe these visual changes [41]. Seizures may result from irritating cerebral edema or from cerebral hemorrhage, particularly with coagulation abnormalities in more severe disease [39].

The long-term prognosis is generally very favorable. Most lesions are thought to be reversible, though imaging shows that some are due to ischemic injury and infarction. It is unknown whether some of these small lesions, subclinical during and shortly after pregnancy, may develop into clinically significant disease later in life. Long-term follow-up of these women is ongoing [42].

Coagulation Abnormalities and HELLP

In about 10–20% of cases of severe pre-eclampsia, thrombotic microangiopathy develops, resulting in the HELLP syndrome. Hemolysis results from shearing of red cells passing through small blood vessels with intima made rough by endothelial damage and fibrin deposition, which also obstructs hepatic sinusoids and causes elevated transaminases [43]. Right upper quadrant pain results from stretching of Glisson's capsule, and if increased pressure is progressive and severe, hepatic rupture can occur.

Normal pregnancy is associated with a relatively hypercoagulable state. Teleologically, this is advantageous to avoid hemorrhage after delivery. In HELLP, hypercoagulability is more extreme in the setting of endothelial damage. Dysfunctional endothelial cells produce altered circulating factors such as von Willebrand factor, thrombomodulin, fibronectin, and PAI-1 [44–47]. Platelets are consumed peripherally through adherence to diffuse sites of endothelial injury. Activated platelets release more thromboxane A2, inhibit prostacyclin production and aggregate, causing further endothelial injury.

Maternal sequelae from HELLP syndrome include disseminated intravascular coagulation, placental abruption, acute renal failure, pulmonary edema and subcapsular hepatic hematoma. Perinatal mortality is approximately 10%, and other complications include preterm birth and sequelae from placental insufficiency. After delivery, hepatic function usually returns to normal within 6 weeks [42].

Pathophysiology and Molecular Mechanisms

Abnormal Placentation

In 1939, E.W. Page proposed that the placenta was central to the pathogenesis of pre-eclampsia [48]. Histological examination of placentas from severe pre-eclampsia shows the classic lesions of acute atherosis with multiple infarcts; sclerotic, narrowed arteries and arterioles; deposition of fibrin and thrombosis. Circulation in the uteroplacental unit is decreased, and vascular resistance increased, with placental ischemia. Further, a placenta is required for pre-eclampsia; a fetus is not, illustrated by pre-eclampsia in molar pregnancies. Conversely, placental removal via delivery resolves pre-eclampsia. Thus, an abnormal placen-

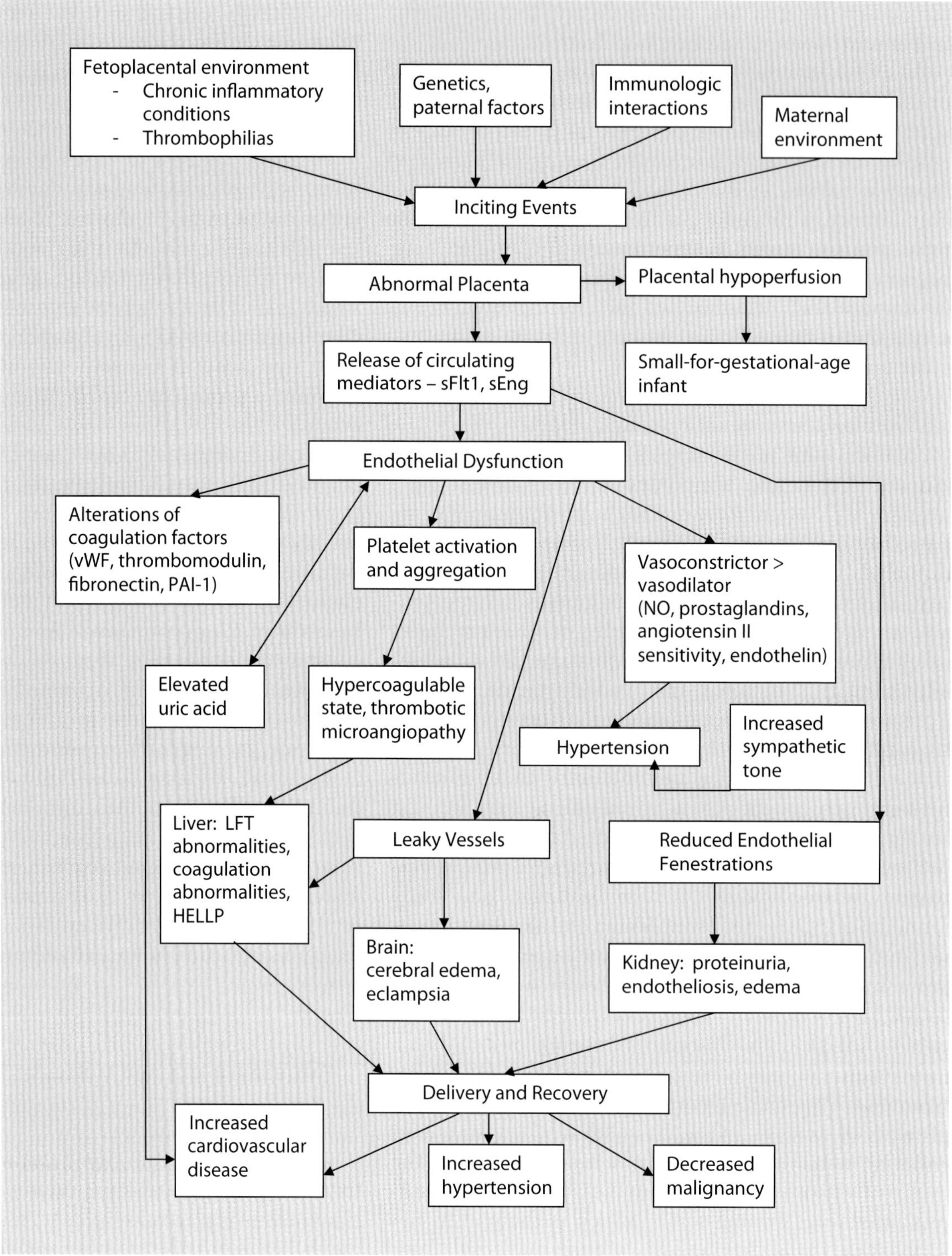

Fig. 1. Summary of the pathogenesis of pre-eclampsia.

68 Sugimoto H, Hamano Y, Charytan D, Cosgrove D, Kieran M, Sudhakar A, Kalluri R: Neutralization of circulating vascular endothelial growth factor (VEGF) by anti-VEGF antibodies and soluble VEGF receptor 1 (sFlt1) induces proteinuria. J Biol Chem 2003;278:12605–12608.
69 Roberts WG, Palade GE: Increased microvascular permeability and endothelial fenestration induced by vascular endothelial growth factor. J Cell Sci 1995;108:2369–2379.
70 Cheifetz S, Bellon T, Cales C, Vera S, Bernabeu C, Massague J, Letarte M: Endoglin is a component of the transforming growth factor-beta receptor system in human endothelial cells. J Biol Chem 1992;267:19027–19030.
71 Gougos A, St Jacques S, Greaves A, O'Connell PJ, d'Apice AJ, Buhring HJ, Bernabeu C, van Mourik JA, Letarte M: Identification of distinct epitopes of endoglin, an RGD-containing glycoprotein of endothelial cells, leukemic cells, and syncytiotrophoblasts. Int Immunol 1992;4:83–92.
72 McAllister KA, Grogg KM, Johnson DW, Gallione CJ, Baldwin MA, Jackson CE, Helmbold EA, Markel DS, McKinnon WC, Murrell J, et al: Endoglin, a TGF-beta binding protein of endothelial cells, is the gene for hereditary haemorrhagic telangiectasia type 1. Nat Genet 1994;8:345–351.
73 Toporsian M, Gros R, Kabir MG, Vera S, Govindaraju K, Eidelman DH, Husain M, Letarte M: A role for endoglin in coupling eNOS activity and regulating vascular tone revealed in hereditary hemorrhagic telangiectasia. Circ Res 2005;96:684–692.
74 Venkatesha S, Toporsian M, Lam C, Hanai J, Mammoto T, Kim YM, Bdolah Y, Lim KH, Yuan HT, Libermann TA, Stillman IE, Roberts D, D'Amore PA, Epstein FH, Sellke FW, Romero R, Sukhatme VP, Letarte M, Karumanchi SA: Soluble endoglin contributes to the pathogenesis of preeclampsia. Nat Med 2006;12:642–649.
75 Levine RJ, Lam C, Qian C, Yu KF, Maynard SE, Sachs BP, Sibai BM, Epstein FH, Romero R, Thadhani R, Karumanchi SA: Soluble endoglin and other circulating antiangiogenic factors in preeclampsia. N Engl J Med 2006; 355:992–1005.
76 Epstein FH: Late vascular effects of toxemia of pregnancy. N Engl J Med 1964;271:391–395.
77 Ray JG, Vermeulen MJ, Schull MJ, Redelmeier DA: Cardiovascular health after maternal placental syndromes (CHAMPS): population-based retrospective cohort study. Lancet 2005;366:1797–1803.
78 Smith GC, Pell JP, Walsh D: Pregnancy complications and maternal risk of ischaemic heart disease: a retrospective cohort study of 129,290 births. Lancet 2001;357:2002–2006.
79 Irgens HU, Reisaeter L, Irgens LM, Lie RT: Long term mortality of mothers and fathers after pre-eclampsia: population based cohort study. Br Med J 2001;323:1213–1217.
80 Cohn BA, Cirillo PM, Christianson RE, van den Berg BJ, Siiteri PK: Placental characteristics and reduced risk of maternal breast cancer. J Natl Cancer Inst 2001;93:1133–1140.
81 Vatten LJ, Romundstad PR, Trichopoulos D, Skjaerven R: Pre-eclampsia in pregnancy and subsequent risk for breast cancer. Br J Cancer 2002;87:971–973.
82 Paltiel O, Friedlander Y, Tiram E, Barchana M, Xue X, Harlap S: Cancer after preeclampsia: follow up of the Jerusalem perinatal study cohort. Br Med J 2004;328:919.
83 Aagaard-Tillery KM, Stoddard GJ, Holmgren C, Lacoursiere DY, Fraser A, Mineau GP, Varner MW: Preeclampsia and subsequent risk of cancer in Utah. Am J Obstet Gynecol 2006;195:691–699.

nephron
Clinical Practice

Nephron Clin Pract 2007;106:c82–c88
DOI: 10.1159/000101802

Published online: June 6, 2007

Alport Syndrome and Thin Basement Membrane Nephropathy

Paul Scott Thorner[a, b]

[a]Division of Pathology, Hospital for Sick Children, and [b]Department of Laboratory Medicine and Pathobiology, University of Toronto, Toronto, Ont., Canada

Key Words
Alport syndrome · Thin basement membrane · Glomerular basement membrane · Type IV collagen

Abstract
Both Alport syndrome and thin basement membrane nephropathy (TBMN) can be considered as genetic diseases of the GBM involving the $\alpha3/\alpha4/\alpha5$ network of type IV collagen. Mutations in any of the *COL4A3*, *COL4A4* or *COL4A5* genes can lead to total or partial loss of this network. Males with mutations in the X-linked *COL4A5* gene develop Alport syndrome with progressive renal disease and sometimes extra-renal disease. Females who are heterozygous for a *COL4A5* mutation are considered to be carriers for X-linked Alport syndrome. Although their clinical course and GBM ultrastructural changes can sometimes mimic TBMN, more often it tends to be more progressive than usually seen in TBMN. Males or females who are heterozygous for *COL4A3* or *COL4A4* mutations usually manifest as TBMN, with non-progressive hematuria, while those who are homozygous or combined heterozygotes develop autosomal-recessive Alport syndrome. Thus, individuals with TBMN can be considered to be carriers for autosomal-recessive Alport syndrome, but there remain some exceptions in which patients heterozygous for *COL4A3* or *COL4A4* mutations develop autosomal-dominant Alport syndrome. Distinguishing between all these groups of patients requires a combination of family history and a renal biopsy for electron microscopic examination of the GBM and immunohistochemical staining of the GBM for the $\alpha3$, $\alpha4$ and $\alpha5$ chains of type IV collagen. Mutational analysis of the *COL4A3*, *COL4A4*, and *COL4A5* genes, whenever it becomes available, will be a valuable adjunct to the diagnostic workup in these patients. Novel therapeutic approaches may one day provide a treatment or cure for these patients, avoiding the need for transplantation and dialysis.

Alport Syndrome

Alport syndrome (also known as hereditary nephritis) is an inherited progressive nephropathy, often associated with sensorineural deafness and ocular lesions. It accounts for 3% of end-stage renal disease (ESRD) in the pediatric population. Alport syndrome arises from mutations in the genes encoding type IV collagen and an understanding of that molecule helps to explain the pathogenesis of this disease [1]. Type IV collagen is the primary structural component of all basement membranes, and Alport syndrome can be viewed as a genetic disease of the basement membrane.

KARGER
Fax +41 61 306 12 34
E-Mail karger@karger.ch
www.karger.com

1660–2110/07/1062–0082$23.50/0

Accessible online at:
www.karger.com/nec

Paul Scott Thorner, MD, PhD
Division of Pathology, The Hospital for Sick Children
555 University Avenue
Toronto, Ont. M5G 1X8 (Canada)
Tel. +1 416 813 5108, Fax +1 416 813 5974, E-Mail paul.thorner@sickkids.ca

nephron
Experimental
Nephrology

nephron
Physiology

nephron
Clinical
Practice

Author Index Vol. 106, No. 2, 2007

KARGER

Fax +41 61 306 12 34
E-Mail karger@karger.ch
www.karger.com

Accessible online at:
www.karger.com/nec

nephron
Experimental
Nephrology

nephron
Physiology

nephron
Clinical
Practice

Subject Index Vol. 106, No. 2, 2007

KARGER

Fax +41 61 306 12 34
E-Mail karger@karger.ch
www.karger.com

Accessible online at:
www.karger.com/nec